CONTRACTING

IN ALL 50 STATES

Licensing Requirements for General and Specialty Contractors
What you need ■ Who to Contact ■ What it'll cost

Compiled & edited by
R. L. Bryson

Craftsman Book Company
6058 Corte del Cedro / P.O. Box 6500
Carlsbad, CA 92018

Library of Congress Cataloging-in-Publication Data

Contracting in All 50 States: who to contact and general licensing
 requirements for every state / compiled & edited by R.L. Bryson.
   p. cm.
   ISBN 1-57218-070-6
 1. Construction industry – Licenses – United States – States – Directories.
 2. Building trades – Licenses – United States – States – Directories.
 3. Contractors – Licenses – United States – States – Directories. I. Bryson, R. L.
HD9715.U52C593   1998
690'.023'73–dc21   98-22680
                                                              CIP

Contents

About This Book

This book will help you, as a construction professional, get information on most initial licensing procedures for the construction trades in all 50 states.

To get this information for you we began by finding out just which agency in each state is responsible for licensing general contractors, residential and commercial contractors, and as many of the trades as possible. We asked about electricians, plumbers, asbestos abatement, HVAC, roofing, etc.

Then we asked each agency the questions you'll need answers to if you want to get a license in the state. Here are some of our questions:

Do I need a license to practice my trade in your state?

Will your state accept the license I already have?

How much will your license cost?

How long is it good for?

Who do I contact to apply for a license?

What do I have to do to qualify for a license?

Do I have to take an exam?

What kind of exam is it and what's it on?

What study materials do I need to prepare for the exam?

There are some licensing requirements basics which we didn't repeat throughout the book. You can pretty much count on having to be 18 or older and have a high school diploma or equivalent to apply for a license anywhere. You'll also usually be asked to furnish proof of your U.S. citizenship or other legal status to work in the U.S.

When a licensing agency wants you to have work experience to take an exam, they'll want you to provide complete documentation of that experience. Anyone who certifies your work experience must also document their license.

You'll need to give information about any license you've ever had in any state and any you still hold. You'll usually be asked to give information about any citations, violations, liens, etc. you've received in construction work. You may also be asked if you've ever filed for bankruptcy or had any criminal charges brought against you.

Most agencies will ask you to supply two passport-sized photos with your application.

We've also included information for those of you who want to look into bidding on Department of Transportation projects. Almost all of the 50 states ask you all about your company organization, personnel and equipment, so they can evaluate your company and prequalify you to bid on projects. We've passed on to you all the information we could get from each state about their process for doing this.

Almost all states also require that any corporation that wasn't originally set up in the state must register with the Secretary of State before doing business in the state. We've included information on where to contact the Secretary of State in all states which require this.

We have left out information on licenses you may need from local agencies. You'll have to check this for each locale you work in. We could only cover state requirements in this book. We also left out information about any taxing agencies you might have to contact to do business in another state. That's probably material for another book.

Finally, a word about accuracy. We've done our best to make all information in this book as up-to-date and accurate as possible. But you may find that things like phone numbers, addresses, and fees have changed since this book was printed. On rare occasions, some requirements for a license may even have changed. All the agencies that handle licensing work spend time and effort modifying their procedures.

And of course, exam dates change every year so we didn't put them in the book at all. But you can still count on this book to do its best to give you a head start in getting through the licensing process.

Alabama

The state of Alabama requires any general contractor working on a commercial or industrial project costing $20,000 or more to get a license. If you're the general contractor on a residential project that costs $10,000 or more, you need a license. Let's look first at the commercial license.

Commercial General Contractor's License

To get a commercial general contractor's license, ask for an application packet from:

Contact

State Licensing Board for General Contractors
400 South Union Street, Suite 235
Montgomery, AL 36130-0201
(334) 242-2839
Fax: (334) 240-3424

On the application you must give the Board:

- complete names and addresses of four references
- confidential financial statement
- proof of insurance
- proof of net worth of at least $10,000

On the application you need to say which of these types of construction work you want to be licensed in:

- building construction
- highways and streets
- municipal and utility
- heavy and railroad construction
- specialty construction

Specialty construction includes work such as:

Mechanical	Landscaping
Electrical	Construction manager
Sheet metal	Fire protection systems
Masonry	Golf courses
Painting	Tennis courts, running tracks and recreational areas
Roofing and siding	
Heating and air conditioning	Sprinkler systems
Bridges	Asbestos material handling and removing
Swimming pools	POL dispensing systems
Demolition	Outdoor advertising signs

Commercial general contractor's license fees: You won't have to take a written exam to get a commercial general contractor's license in Alabama. It will cost you $300 to file an application for a license. All licenses expire December 31 each year.

Alabama doesn't accept a license in another state for this type of license.

Residential General Contractor's License

If you want to bid and work on residential projects, you need to be licensed by a different agency. It's called the:

Homebuilders Licensure Board
400 South Union Street, Suite 195
Montgomery, AL 36130-3605
(334) 242-2230
Fax: (334) 263-1397

When you apply to this Board you must provide:

- complete names and addresses of two references
- a financial statement
- a check payable to the Holloway Credit Bureau Company for a credit report ($20 for individuals, $45 for corporation or partnership)

To get the residential contractor's license, you have to pass a six-hour exam given by Block and Associates of Gainesville, Florida. For information on the exam, contact:

Block and Associates
2100 NW 53rd Avenue
Gainesville, FL 32653
(800) 280-EXAM

The exam is open book, on the following topics:

Subject	Number of questions
Concrete	9 - 11
Rough carpentry	19 - 21
Interior/exterior finish	17 - 19
Roofing	1 - 3
Excavation	9 - 11
Plans and specifications	9 - 11
Contract management	3 - 5
Personnel regulations	3 - 5
Project management	2 - 4
Financial management	1 - 3
Tax laws	1 - 3
Licensing laws	2 - 4
Lien laws	0 - 2
Safety - OSHA	0 - 2

Residential general contractor's license fees: The license costs $180 and it expires December 31.

Recommended Reading for the Residential Contractor's Exam

- 📖 *Alabama Home Builder's Reference Manual*, 1995, Block and Associates, 2100 NW 53rd Ave., Gainesville, FL 32653

- 📖 *Alabama Home Builder Digest*, 1995, Block and Associates, 2100 NW 53rd Ave., Gainesville, FL 32653

- 📖 *Standard Building Code*, 1994

- 📖 *CABO, One and Two Family Dwelling Code*, 1995, Southern Building Code Congress International, Birmingham, AL

- 📖 *Span Tables for Joists and Rafters*, 1993 with 1992 supplements, American Forest and Paper Association, Washington, DC

- 📖 *Carpentry*, Gaspar Lewis, 1995, Delmar Publishers, P.O. Box 6904, Florence, KY 41022

- 📖 *Gypsum*, Gypsum Association, Washington, DC

Electrician's Licenses

To do electrical work throughout the state of Alabama you need a license. To apply for an electrician's license, contact:

Contact

Alabama Electrical Contractors Board
660 Adams Avenue, Suite 254
P. O. Box 56
Montgomery, AL 36101-0056
(334) 263-3407

The Board issues electrical contractor's and journeyman's licenses.

Electrical Contractor's License

To qualify for a contractor's license you need four years' work experience in designing, planning, laying out and supervising electrical construction, and installing electrical components. You can substitute up to two years of education in a Board-approved electrical curriculum at the rate of one year of education for a half year of work.

Alabama requires you to pass an exam to get a license. The exam is seven hours long — four hours in the morning and three hours in the afternoon. Here's the content of the exam:

Part I *(closed book, one hour, 50 questions) — 17% of total grade:*

Subject	Number of questions
General theory	24 - 26
NEC chapter 1	6 - 8
NEC chapter 2	1 - 3
NEC chapter 3	0 - 2
NEC Article 90	0 - 2
NEC chapter 4	6 - 8
NEC chapter 5	2 - 4
NEC chapter 6	2 - 4

Part II *(open book, three hours, 100 questions) — 33% of total grade:*

Subject	Number of questions
General theory	8 - 10
NEC chapter 1	1 - 3
NEC chapter 2	9 - 11
NEC chapter 3	12 - 14
NEC chapter 4	16 - 18
NEC chapter 5	4 - 6

Subject	Number of questions
NEC chapter 6	7 - 9
NEC chapter 7	0 - 2
NEC chapter 8	1 - 3
NEC chapter 9	2 - 4
Business	29 - 31

Part III (open book, three hours, 30 questions) — 50% of total grade:

Subject	Number of questions
Service	4 - 6
Voltage drop	2 - 4
3Ø 208V delta and wye loads	2 - 4
Conduit fill	3 - 5
Conductor ampacity	2 - 4
Motors (branch circuit)	0 - 2
Motors (feeders)	0 - 2
Motors (protection)	0 - 2
Motors (control)	2 - 4
Appliance loads	1 - 3
Transformers, 3Ø transformer calculation	1 - 3
Grounding conductor and equipment ground	0 - 2
Power factor, power, and volt-amps	0 - 2

Electrical contractor's license fees: It will cost you $150 nonrefundable to take the exam and another $200 for the contractor's license.

If you have an unrestricted electrical contractor's license from Florida, Georgia, North Carolina, South Carolina or Mississippi, you can get a license from the Board by reciprocity. You must have passed their exam with a grade of at least 75 to get the license and hold it in good standing. You will still have to pay the $350 fee for the Alabama license. If you got your electrical contractor's license in Mississippi, you'll have to take the business section of the Alabama electrical contractor's exam and pay $28 for that part of the exam as well.

Journeyman Electrician's License

To get a journeyman's license you have to pass the Alabama journeyman's exam. To qualify for the exam you need four years of electrical work experience. You can substitute up to two years of education in electrical studies for one year of the work experience.

The exam is six hours long — three hours in the morning and three hours in the afternoon. Here's the content of the exam:

Part I (closed book, one hour, 50 questions)— 25% of total grade:

Subject	Number of questions
General theory	17 - 19
Materials	1 - 3
Field application	9 - 11
NEC chapter 1 Articles 100, 110	8 - 10
NEC chapter 2	4 - 6
NEC chapter 3	3 - 5
NEC chapter 4	0 - 2
NEC chapter 9	0 - 2

Part II (open book, two hours, 50 questions) — 25% of total grade:

Subject	Number of questions
General theory	3 - 5
Materials	1 - 3
NEC chapter 1	3 - 5
NEC chapter 2	17 - 19
NEC chapter 3	12 - 14
NEC chapter 4	5 - 7
NEC chapter 5	0 - 2
NEC chapter 6	0 - 2
NEC Article 90	0 - 2

Part III (open book, three hours, 30 questions) — 50% of total grade:

Subject	Number of questions
Residential service	4 - 6
Conduit fill	2 - 4
Motors	3 - 5
Ambient temperature	3 - 5
Efficiency, power factor, neutral loads	0 - 2
Box fill	0 - 2
Transformers	3 - 5
Voltage drop	3 - 5
Conductor ampacity	1 - 3
Appliance loads	1 - 3

 Journeyman's license fee: It will cost you $100 nonrefundable for the exam and $35 for the journeyman's license.

Recommended Reading for the Electrical Exams

Electrical contractor's exam:

📖 *Contractor's Reference Manual,* Alabama Edition, 1997, Block & Associates, 2100 NW 53rd Ave., Gainesville, FL 32653

📖 *NFPA 70 - National Electrical Code,* 1996 edition, National Fire Protection Association, 11 Tracey Ave., Avon, MA 02322

📖 *American Electricians Handbook,* 1996, 13th edition, Croft/Summers, McGraw-Hill Inc., Box 543, Blacklick, OH 43004-0543

📖 *Electrical Review for Electricians,* 9th edition, 1996, J. Morris Trimmer and Charles Pardue, Construction Bookstore, 100 Enterprise Place, Dover, DE 19903-7029

Journeyman exam:

📖 *NFPA 70 - National Electrical Code,* 1996 edition, National Fire Protection Association, 11 Tracey Ave., Avon, MA 02322

📖 *American Electricians Handbook,* 1996, 13th edition, Croft/Summers, McGraw-Hill Inc., Box 543, Blacklick, OH 43004-0543

📖 *Electrical Review for Electricians,* 9th edition, 1996, J. Morris Trimmer and Charles Pardue, Construction Bookstore, 100 Enterprise Place, Dover, DE 19903-7029

You can get these books from:

Builders' Book Depot
1033 East Jefferson, Suite 500
Phoenix, AZ 85034
(800) 284-3434

HVAC Contractor's Certification

To protect the public and identify knowledgeable and capable contractors, Alabama has a state board that certifies HVAC contractors. To apply for a certificate, contact:

Contact

Board of Heating and Air Conditioning Contractors
100 North Union Street
Montgomery, AL 36130
(334) 242-5550

Fee

HVAC contractors certification fees: It will cost you $75 nonrefundable to file an application for certification. The Board may also require you to post a $5,000 bond.

Certification is good for one year but you must renew it within 90 days after October 1 each year.

You'll have to pass a two-part exam to be certified. Each part is open book and three hours long with 60 multiple choice questions. Here's the content of the exam:

Part I:

Subject	Number of questions
Code compliance (Standard Mechanical Code)	15 - 25
Code compliance (Standard Gas Code)	2 - 4
HARV (general)	20 - 30
HARV (load calculation)	10 - 15
Ducts (safety and fire prevention)	4 - 6
Ducts (vapor removal and hood systems)	1 - 3

Part II:

Subject	Number of questions
HARV (maintenance)	10 - 15
HARV (controls)	7 - 15
Ducts	6 - 10
Safety (refrigeration)	8 - 13
Insulation	4 - 7
Psychrometric analysis	4 - 7
Sheet metal	4 - 8
Ducts, construction (fiberglass)	4 - 8

Recommended Reading for HVAC Certification

Standard Mechanical Code 1994 and Standard Gas Code 1994, Southern Building Code Congress International, Inc., 900 Montclair Road, Birmingham, AL 35213-1206

ANSI/ASHRAE 15-94 - Safety Code for Mechanical Refrigeration, 1994 ANSI, 11 West 42nd St., New York, NY 10036

Refrigeration and Air Conditioning, 1995, 3rd edition, ARI, Prentice Hall, P.O. Box 11071, Des Moines, IA 50336-1071

Manual N - Load Calculation for Commercial Summer and Winter Air Conditioning, 1988, 4th edition, Air Conditioning Contractors of America, 1712 New Hampshire Ave., NW, Washington, DC 20009

Manual J - Load Calculation for Residential Winter and Summer Air Conditioning, 1986, 7th edition, Air Conditioning Contractors of America, 1712 New Hampshire Ave., NW, Washington, DC 20009

Recommended Reading for HVAC Certification (continued)

📖 *Trane Ductulator*, 1976, Trane Company, 8929 Western Way, Suite #1, Jacksonville, FL 32256

📖 *NFPA 90A - Installation of Air Conditioning and Ventilating Systems*, 1993, National Fire Protection Association, 1 Batterymarch Park, Box 9101, Quincy, MA 02269-9101

📖 *NFPA 90B - Installation of Warm Air Heating and Air Conditioning Systems*, 1993, National Fire Protection Association, 1 Batterymarch Park, Box 9101, Quincy, MA 02269-9101

📖 *NFPA 96 - Standard for Ventilation Control and Fire Protection of Commercial Cooking Operations*, 1994, National Fire Protection Association, 1 Batterymarch Park, Box 9101, Quincy, MA 02269-9101

📖 *HVAC Duct Construction Standards - Metal and Flexible*, 1995, Sheet Metal and Air Conditioning Contractor's National Assoc., Box 221230, Chantilly, VA 22022-1230

📖 *Fibrous Glass Duct Construction Standards, 1993*, 2nd edition, North American Insulation Manufacturers Assoc., 44 Canal Center Plaza, Suite #310, Alexandria, VA 22314

Plumber's Licenses

You need a license to do plumbing work in Alabama. To get an application for a license, contact:

Alabama Plumbers and Gas Fitters Examining Board
11 West Oxmoor Road, Suite 104
Birmingham, AL 35209
(205) 945-4857

Master's and journeyman plumber's exam fees: The master plumber's exam costs $100, the journeyman's $50, and both are nonrefundable. If you pass the exam, the exam fee covers your license fee for one year. Renewal is $100 for masters and $25 for journeymen. Licenses expire on December 31.

Alabama also registers apprentice plumbers. The fee is $25. Apprentices can only work under the supervision of journeymen or masters.

Master Plumber's License

The Board issues master and journeyman licenses and each requires you to pass an exam. You need one year of work experience as a journeyman plumber to qualify for the master plumber's exam. The exam has three parts:

Part I *(closed book, 100 questions, two hours, 40% of grade):*

Subject	Number of questions
Administration	8 - 12
Basic principles	1 - 3
Definitions	4 - 6
General regulations	10 - 12
Materials	3 - 5
Joints and connections	2 - 4
Traps	3 - 4
Cleanouts	3 - 4
Interceptors	4 - 8
Fixtures	8 - 12
Hangers and supports	2 - 6
Indirect wastes	4 - 8
Water supply and distribution	8 - 12
Drainage systems	8 - 12
Vents and venting	8 - 12
Storm drains	1 - 3

Part II *(open book, 30 questions, one hour, 20% of grade):*

Subject	Number of questions
Fitting identification	8 - 12
General knowledge	6 - 10
Developed length	5 - 10
Drainage calculations	1 - 3
Vent calculations	2 - 4

Part III *(open book, 40 questions, three hours, 40% of grade):*

Subject	Number of questions
Isometric analysis	40

Journeyman Plumber's License

To take the journeyman plumber's exam you need to complete a Board-approved apprenticeship program or work two years as an apprentice. Here's the content of the exam:

Part I (closed book, 100 questions, three hours, 50% of grade):

Subject	Number of questions
Administration	6 - 10
Basic principles	1 - 3
Definitions	6 - 8
General regulations	6 - 8
Materials	4 - 6
Joints and connections	2 - 4
Traps	3 - 4
Cleanouts	3 - 4
Interceptors	4 - 6
Fixtures	4 - 6
Hangers and supports	4 - 6
Indirect wastes	1 - 3
Water supply and distribution	6 - 10
Drainage systems	8 - 12
Vents and venting	8 - 12
Storm drains	1 - 3
Fitting identification	12 - 18

Part II (open book, 50 questions, three hours, 50% of grade):

Subject	Number of questions
General knowledge	10 - 16
Developed length	6 - 12
Isometric analysis	25 - 30

Recommended Reading for the Plumbing Exams

📖 *Standard Plumbing Code 1994,* Southern Building Code Congress International, Inc., 900 Montclair Road, Birmingham, AL 35213-1206

📖 *Plumbing Technology: Design and Installation,* 1994, 2nd edition, Lee Smith, Delmar Publishers, P.O. Box 6904, Florence, KY 41022

📖 *Plumbing,* L. V. Ripka, American Technical Publishers, 1155 West 175th Street, Homewood, IL 60403

Gas Fitter's Licenses

You need a license to do gas fitting work in Alabama. To get an application for a license, contact:

Alabama Plumbers and Gas Fitters Examining Board
11 West Oxmoor Road, Suite 104
Birmingham, AL 35209
(205) 945-4857

The Board issues master's and journeyman's licenses and each requires you to pass an exam. You need one year of work experience as a journeyman gas fitter to qualify for the master gas fitter exam. The exam lasts three hours. It's open book with 60 questions. Here's an overview of the master gas fitter's exam:

Subject	Number of questions
Code compliance	15 - 25
Fuel gas systems	20 - 30
HARV controls	5 - 10
Piping	2 - 8

To take the journeyman gas fitter's exam you need to complete a Board-approved apprenticeship program or work two years as an apprentice. The exam lasts three hours. It's open book with 50 questions. It includes:

Subject	Number of questions
Code compliance	12 - 20
Fuel gas systems	20 - 30
HARV controls	5 - 15
Piping	2 - 8

Master and journeyman gas fitter's license fees: The master gas fitter's exam costs $100, the journeyman $50, and both are nonrefundable. If you pass the exam, the exam fee covers your license fee for one year. Renewal is $100 for master gas fitters and $25 for journeymen gas fitters. Licenses expire on December 31.

Alabama also registers apprentice gas fitters. The fee is $25. Apprentices can only work under the supervision of journeymen or masters.

Recommended Reading for the Gas Fitter's Exams

📖 *Standard Gas Code 1994,* Southern Building Code Congress International, Inc., 900 Montclair Road, Birmingham, AL 35213-1206

📖 *NFPA 54 - National Fuel Gas Code,* 1996, National Fire Protection Association, 11 Tracey Ave., Avon, MA 02322

Recommended Reading for the Gas Fitter's Exams (continued)

📖 *Pipefitters Handbook,* 1967, 3rd edition, Forest Lindsey, Industrial Press, Inc., 200 Madison Avenue, New York, NY 10016

📖 *Refrigeration and Air Conditioning,* 1995, 3rd edition, ARI, Prentice Hall, P.O. Box 11071, Des Moines, IA 50336-1071

📖 *Fundamentals Six Pack Special,* 1990, American Gas Association, 1515 Wilson Blvd., Arlington, VA 22209

Department of Transportation (DOT)

To bid on Alabama Department of Transportation projects you have to be prequalified by the Department. To get an application for prequalification, contact:

Contact

Alabama Department of Transportation
Prequalification Engineer, Room E-101
1409 Coliseum Blvd.
Montgomery, AL 36130-3050
(334) 242-6059

Some of the details the Department will ask you about are your organization, personnel, financial condition, equipment, and experience. You'll be asked what type of work you want to be qualified for. Here are the types of work the Department uses:

Major bridges
Intermediate bridges
Minor bridges
Movable bridge rehabilitation
Grading
Flexible paving
Portland cement concrete paving
Major drainage
Minor drainage

The following types of work are included under the category of Specialty work:

- electrical
- traffic control signing and maintenance
- permanent roadway signing
- pavement markings
- guardrails
- fencing

- demolition
- piling, furnishing and driving
- environmental controls
- buildings and service stations
- earth retention systems
- drill and blast

- bridge painting
- tunnel maintenance
- grassing, seeding and sodding, landscaping
- bridge repair and bridge deck repair
- mowing, maintenance
- utilities
- pavement repair
- dredging
- marine
- other

Unless the Alabama State Licensing Board for General Contractors exempts you, you need a license from them to bid on state projects. On any project with federal participation you'll need the license before you can be awarded the contract. If you want to do any mowing work for the Department you need to get a license from the Board first.

Prequalification is good for one year, but the Department reserves the right to ask you for another prequalification statement at any time.

Out-of-State Corporations

All corporations must register to do business in Alabama with the Office of the Secretary of State. Their address is:

Contact

Office of the Secretary of State
P.O. Box 5616
Montgomery, AL 36103
(334) 242-5324
Fax: (334) 240-3138

Alaska

According to Alaska law you must be registered with the Department of Commerce and Economic Development to bid or work as a contractor in the state. Alaska licenses these types of contractors:

General (excluding residential construction)
General (including residential construction)
Mechanical
Specialty

Specialty contractors include the following:

Access flooring
Acoustical and insulation
Asbestos abatement
Carpentry, finish
Carpentry, rough
Communications
Concrete and paving
Demolition
Drilling
Drywall
Electrical
Elevator and conveying system
Excavation
Fence and guardrail
Floor covering
Glazing
Landscaping
Liquid or gas storage tank
Low voltage alarm and signal device

Marine
Masonry
Mechanical, exempt
Painting
Plaster
Road construction
Roofing
Security system
Sheet metal
Sign
Solid fuel appliance
Steel erection
Tile and terrazzo
Wallcovering
Water and sewer
Water system
Welding
Other

You can be licensed as a specialty contractor in only three distinct nonmechanical trades from this list.

Nonresidential General Contractor's License

To apply for a general contractor license which doesn't let you do residential work, contact:

Contact

Department of Commerce and Economic Development
Division of Occupational Licensing
333 Willoughby Avenue, 9th Floor
P.O. Box 110806
Juneau, AK 99801-0800
(907) 465-2546

On the application, you will be asked for:

- proof of liability insurance — at least $20,000 property damage, $50,000 injury or death to one person, $100,000 injury or death to more than one person
- proof of workers' compensation insurance
- surety bond — $10,000 for a general contractor or $5,000 for a specialty or mechanical contractor

Fee

Nonresidential general contractor's license fees: Alaska doesn't require an exam for the nonresidential type of license. It will cost you $190 to register — $50 nonrefundable for the application and $140 for the license. The license is good for two years.

Residential General Contractor's License

There's quite a bit more involved in getting a general contractor's license that lets you work on residential construction. First you must successfully complete the Alaska Craftsman Home Program (ACHP) or its equivalent, or a post-secondary course in Arctic engineering or its equivalent. You have to complete the course within two years before you apply for the license. ACHP is a private, nonprofit organization which holds workshops around the state. You can contact the ACHP at:

Contact

ACHP
900 West Fireweed Lane, Suite 201
Anchorage, AK 99503
(907) 258-2247 / (800) 699-9276 in Alaska
Fax: (907) 258-5352

After completing the course, the next step is to get an Endorsement Application for Residential Construction from:

Contact

Department of Commerce and Economic Development
Division of Occupational Licensing
333 Willoughby Avenue, 9th Floor
P.O. Box 110806
Juneau, AK 99811-0806
(907) 465-3035

Fee

Residential general contractor's license fees: They'll send you an application and information on their Residential Contractor Endorsement Examination. The application will cost you $50 nonrefundable, the endorsement $50, and the exam $75. This exam is given every three months in Anchorage, Fairbanks, and Juneau. You have to apply at least 45 days before the exam date. The four-hour exam is open book, with 100 multiple choice questions. Exam content includes:

Subject	Percent of exam
Excavation & site preparation	6
Windows & doors	6
Building science	8
Finish work	9
Insulation	10
Ventilation & heating	12
Roofing & attics	15
Concrete & foundation system	16
Rough carpentry & walls	18

You will also need to apply for a business license from the Business Licensing Section. You can contact them at:

Contact

Department of Commerce and Economic Development
Business Licensing Section
333 Willoughby Avenue, 9th Floor
P.O. Box 110806
Juneau, AK 99811-0806
(907) 465-2550

The license will cost you $50 nonrefundable and it's good for two calendar years. If you'll be registering as a corporation, you need to contact:

Contact

Corporations Section
Division of Banking, Securities and Corporations
Department of Commerce and Economic Development
P.O. Box 110807
Juneau, AK 99811-0807
(907) 465-2530

Recommended Reading for the Residential Contractor Exam

📖 *Design and Control of Concrete Mixtures,* 1988/1990, 13th edition, Portland Cement Association, 5420 Old Orchard Road, Skokie, IL 60077-1083

📖 *Excavation & Grading Handbook,* 1991, Nicholas E. Capachi, Craftsman Book Company, 6058 Corte del Cedro, P.O. Box 6500, Carlsbad, CA 92018

**Recommended Reading for the
Residential Contractor Exam (continued)**

📖 *Modern Carpentry,* 1992, Willis H. Wagner, The Goodheart-Willcox Company, Inc.

📖 *Carpentry and Building Construction,* 1993, Feirer, Hutchings & Feirer, McGraw-Hill Inc., Box 543, Blacklick, OH 43004-0543

📖 *Alaska Craftsman Home Building Manual,* Alaska Craftsman Home Program

📖 *Using Gypsum Board for Walls and Ceilings,* 1985, Gypsum Association

📖 *Uniform Building Code,* 1991, International Conference of Building Officials

You can get these books from:

Builders' Book Depot
1033 Jefferson, Suite 500
Phoenix, AZ 85034
(800) 284-3434

Certificates of Fitness for the Trades

The Alaska Department of Labor requires a Certificate of Fitness for plumbers, electricians, and workers with asbestos abatement, hazardous paint, explosives, and boilers. Here's a summary of the requirements:

Journeyman plumber

Required training or experience: 8000 hours work experience

Fee (nonrefundable): $210

Exam: yes

Certificate good for 2 years

Restricted plumber

Required training or experience: 4000 hours work experience

Fee (nonrefundable): $210

Exam: yes

Certificate good for 2 years

Trainee plumber

Required training or experience: None

Fee (nonrefundable): $210

Exam: no

Certificate good for 2 years

Journeyman electrician

Required training or experience: 8000 hours work experience

Fee (nonrefundable): $210

Exam: yes

Certificate good for 2 years

Residential electrician

Required training or experience: 4000 hours work experience

Fee (nonrefundable): $210

Exam: yes

Certificate good for 2 years

Trainee electrician

Required training or experience: None

Fee (nonrefundable): $210

Exam: no

Certificate good for 2 years

Journeyman lineman

Required training or experience: 8000 hours work experience

Fee (nonrefundable): $210

Exam: yes

Certificate good for 2 years

Lineman trainee

Required training or experience: none

Fee (nonrefundable): $210

Exam: no

Certificate good for 2 years

Reciprocal journeyman electrician

Required training or experience: none

Fee (nonrefundable): $210

Exam: no

Certificate good for 2 years

Asbestos abatement

Required training or experience: 40 hours training

Fee (nonrefundable): $100

Exam: no

Certificate good for 2 years

Hazardous paint handler

Required training or experience: 16 hours training

Fee (nonrefundable): $100

Exam: no

Certificate good for 3 years

Explosives handler

Required training or experience: 6 months work experience

Fee (nonrefundable): $100

Exam: yes

Certificate good for 3 years

Boiler operator class 1

Required training or experience: 2 years work experience or class 2 license 1 year

Fee (nonrefundable): none

Exam: yes

Certificate good for 3 years

Boiler operator class 2

Required training or experience: 1 year work experience or class 3 license 6 months

Fee (nonrefundable): none

Exam: yes

Certificate good for 3 years

Boiler operator class 3

Required training or experience: 6 months work experience

Fee (nonrefundable): none

Exam: yes

Certificate good for 3 years

Boiler operator class 4

Required training or experience: none

Fee (nonrefundable): none

Exam: no

Certificate good for 3 years

Study guides for electricians' certificates are the *1996 National Electrical Code, Lineman and Cableman's Handbook,* and the *1993 National Electrical Safety Code.* Study guide for plumbers' certificates is the *1994 Uniform Plumbing Code.*

To apply for a Certificate of Fitness, contact:

Alaska Department of Labor/Mechanical Inspection
3301 Eagle Street #302
P.O. Box 107020
Anchorage, AK 99510-7020
(907) 269-4925
E-mail Brandi_Barger@commerce.state.ak.us

Department of Transportation (DOT)

You don't need to be prequalified to bid on Alaska Department of Transportation projects.

Out-of-State Corporations

Out-of-state corporations must get a Certificate of Authority to do business in Alaska from the Alaska Secretary of State. To apply for this certificate, contact:

Corporations Section
Department of Commerce and Economic Development
P.O. Box 110808
Juneau, AK 99811-0808
(907) 465-2530
Fax: (907) 465-3257

Arizona

Basically, you need a license to bid on any job over $750 in Arizona.

Arizona issues separate licenses for commercial and residential work for each particular trade or construction field. Licenses are good for two years. To apply for either license, contact:

Contact

Arizona Registrar of Contractors
800 West Washington
6th Floor
Phoenix, AZ 85007-2940
(602) 542-1525

You'll have to pass an exam before you can apply for a license in Arizona. To apply for the exam, contact:

Contact

A.C.S.I., Inc.
National Assessment Institute
1033 East Jefferson
Phoenix, AZ 85034
(602) 258-2143

You must pass the exam within six months from when you apply for it. After that you'll have one year to apply for your license. The exam allows two and one-half hours for a business management section on Arizona contractor licensing laws and regulations, construction project management, and business and financial management. Everyone must take this part of the exam.

Then there's a three-hour exam (except for the *Electrical* and *Electrical and transmission lines* exams which are four hours long) on a trade you pick.

Here's a summary of the commercial construction trades you can be licensed in, years of experience you need, and whether you have to pass a trade exam for the license:

Trade	Years experience required	Exam required?
General engineering	4	yes
Blasting	4	yes

Trade	Years experience required	Exam required?
General drilling	4	yes
Excavating, grading & oil surfacing	4	no
Piers & foundations	4	yes
Swimming pools	4	yes
Steel & aluminum erection	4	yes
Sewers, drains & pipe laying	4	yes
Asphalt paving	4	no
Seal coating of parking lots, etc.	2	no
Waterworks	4	yes
Electrical & transmission lines	4	yes
Swimming pools, including solar	4	yes
Landscaping & irrigation systems	4	yes
General commercial	4	yes
General small commercial	4	yes
Acoustical systems	2	no
Awnings, canopies, carports/patio covers	2	no
Boilers, steamfitting/process piping	4	yes
Swimming pool service & repair	1	no
Carpentry	4	yes
Floor covering	2	no
Concrete	4	yes
Drywall	2	yes
Electrical	4	yes
Elevators	4	yes
Carpets	2	no
Fencing	3	no
Fire protection systems	4	yes
Ornamental metals	2	no
Landscaping	2	no
Lightweight partitions	2	no
Masonry	4	no
Painting & wall covering	2	no
Plastering	3	no
Plumbing	4	yes
Signs	3	no
Air conditioning & refrigeration	4	yes
Insulation	2	no
Septic tanks and systems	3	yes
Roofing	4	yes
Irrigation systems	2	yes
Sheet metal	2	no
Ceramic, plastic, & metal tile	4	yes

Trade	Years experience required	Exam required?
Commercial, industrial refrigeration	4	yes
Water well drilling	2	yes
Water conditioning equipment	2	yes
Welding	2	no
Wrecking	3	no
Comfort heating, ventilating, evaporation cooling	2	yes
Finish carpentry	2	no
Carpentry, remodeling & repairs	4	yes
Steel reinforcing bar & wire mesh	4	yes
Appliances	2	no
Wood floor laying & finishing	2	no
Glazing	3	yes
Low voltage communications systems	2	yes
Boilers, steamfitting, process piping including solar	4	yes
Plumbing including solar	4	yes
Solar plumbing — liquid systems only	½	yes
Air conditioning, refrigeration including solar	4	yes

Here's the information for the residential construction trades:

Trade	Years experience required	Exam required?
General building	4	yes
General remodeling & repair	4	yes
General engineering	4	yes
General swimming pool	4	yes
Factory fabricated pools & accessories	2	yes
General swimming pool (solar)	4*	yes
Pre-manufactured spas & hot tubs	2	no
Acoustical systems	1	no
Excavating, grading & oil surfacing	3	no
Awnings & canopies	1	no
Boilers including solar	4	yes
Boilers	4	yes
Swimming pool service & repair	1	no
Carpentry	4	yes
Floor covering	2	no
Wood flooring	1	no
Carpet	1	no
Concrete	4	yes
Gunite & shotcrete	3	no
Terrazzo	3	no

*3½ years practical experience plus ½ year solar experience

Trade	Years experience required	Exam required?
Drywall	2	no
Electrical	4	yes
Low voltage communication systems	1	yes
Asphalt paving	4	no
Asphalt coating & parking appurtenances	2	no
Fencing	3	no
Fencing other than masonry	1	no
Blasting	4	yes
Fire protection	4	yes
Structural steel & aluminum	4	yes
Welding	1	no
Ornamental metals	1	no
Rebar & wire mesh	2	no
Elevators	4	no
Landscaping & irrigation systems	3	yes
Landscaping	2	no
Irrigation systems	2	yes
House moving	2	no
Finish carpentry	2	no
Cultured marble	1	no
Weatherstripping	1	no
Masonry	4	yes
Painting & wall covering	2	no
Surface preparation & waterproofing	1	no
Plastering	3	no
Swimming pool plastering	2	no
Lathing	2	no
Plumbing including solar	4	yes
Plumbing	4	yes
Gas piping	2	yes
Sewers, drains & pipe laying	4	yes
Water conditioning equipment	2	yes
Solar plumbing (liquid systems only)	½	yes
Signs	1	no
Air conditioning & refrigeration including solar	4	yes
Air conditioning & refrigeration	4	yes
Warm air heating, evaporative cooling & ventilating	2	yes
Evaporative cooling & ventilators	2	no
Insulation	2	no
Foam insulation	1	no
Sewage treatment systems	3	yes
Precast waste treatment systems	2	yes
Roofing	4	yes
Foam & foam panel roofing	2	no

Trade	Years experience required	Exam required?
Roofing shingles & shakes	2	no
Sheet metal	2	no
Ceramic, plastic & metal tile	3	no
Swimming pool tile	2	no
Drilling	2	yes
Limited remodeling & repair	4	yes
Minor home improvements	0	no
Appliances	2	no
Glazing	3	no
Mobile home remodeling & repair	4	yes
Wrecking	3	no

Fee

Commercial and residential license fees: All exam questions are multiple choice. The cost for the exam is $85 for the business law part or one trade exam. If you apply for the business law and one trade exam, the cost is $125.

National Assessment Institute has Study Guides for each exam. When they get your application for the exam, they'll mail you guides for the exams you're taking.

After you pass the exam, you can complete your license application. You'll need a financial statement, proof of Workers' Compensation Insurance if you have employees, and a bond. The amount of the bond depends on the type of license you're getting and the gross amount of business you expect to do in Arizona. For commercial contractor licenses, bonds can be from $2,500 to $90,000 for each license. For residential contractor licenses, they're from $1,000 to $15,000 each. Residential contractors also need to pay into the Arizona Recovery Fund. The amount of the payment may vary from year to year depending on how much the Fund has accumulated over the year.

License fees are:

$920 for each commercial construction general or engineering classification

$670 for each commercial construction trade

$460 for each residential construction general or engineering classification

$335 for most residential construction trades

Arizona has reciprocity agreements with California, Nevada, and Utah which will let you skip any trade exam you already have a license for. To qualify for this you must:

- have been licensed in good standing for five years or more
- have passed an equivalent trade exam (except for electrical and plumbing trade exams from Nevada)
- send a license verification with your application
- pass the Arizona business management exam

Your experience and licensing in another state may help you qualify for a license but you still must be issued a license by the Arizona Registrar of Contractors.

Department of Transportation (DOT)

To bid on highway projects in Arizona, you need a commercial construction license and prequalification with the Arizona Department of Transportation. Their address is:

Contact

Arizona Department of Transportation
Highways Division
Contracts and Specifications Services
1651 West Jackson Street, Room 121-F
Phoenix, AZ 85007-3217
(602) 255-7221

Their application will ask you:

- what type of work you want to prequalify for
- what experience you have
- what equipment you own
- what your financial condition is

You must file the application at least 15 days before the opening bid date for any project you're bidding on. Prequalification is good for 15 months from the date of the financial statement you send with your application.

The Department of Transportation will look at your application and classify your company as:

Classification	Basis	Cost limit of job you can prequalify for
Inexperienced	no experience as a prime highway contractor	$300,000
New	new business with experience in highway construction with other prime contracting businesses	five times the net worth of your company
Unknown	experience as a prime contractor but not in highway construction as a prime contractor with DOT	the larger of five times the net worth of your company or the cost of the largest highway project you've completed as a prime contractor for any other public agency
Known	experience as at least one highway construction project as a prime contractor for the DOT	ten times the net worth of your company or unlimited if ten times the net worth of your company is more than $30,000,000

Out-of-State Corporations

If your corporation isn't chartered in Arizona, you'll also need a letter from the Arizona Corporation Commission authorizing you to do business in Arizona. Their address is:

Contact

Arizona Corporation Commission
1300 West Washington
Phoenix, AZ 85007
(602) 542-3026

Arkansas

To bid and work on construction projects in Arkansas that cost $20,000 or more, you must get a contractor's license. To apply for the license, contact:

Contact

Contractors Licensing Board
621 East Capitol
Little Rock, AR 72202
(501) 372-4661

The application will ask you for:

- a financial statement
- a statement of your experience
- professional references

Contractor's Classifications

You'll also be asked which classifications of work you want to do in the state. Here's an outline of the classifications Arkansas uses:

Marine

Tunnels and shafts

Energy and power plants

Dams, dikes, levees, canals

Mining surface and underground

Oil field construction

Oil refineries

Highway, railroad and airport construction:

    grading, drainage
    base, paving
    bridges, culverts

    railroad construction
    miscellaneous and specialty items

Municipal and utility construction:

    underground piping
    water and sewer plants

    grading, drainage streets and roads
    base and paving

Building contracting:

acoustical treatments	elevators, escalators
carpentry, framing, millwork	erection, fabrication of
drywall	structural steel
floor covering	concrete
foundations	sheet metal
glass, window, door construction	roofing
institution and recreational equipment	conveyors
lathe, plaster, stucco	sandblasting
masonry	golf courses
ornamental and miscellaneous metal	tennis courts
painting, interior decorating	swimming pools
roof decks	outdoor advertising
site and subdivision development	excavation
special coatings, waterproofing	landscaping
tile, terrazzo, marble	fencing
insulation	millwright

Mechanical contracting:

plumbing	pollution control
process piping	pneumatic tube systems
HVAC	temperature controls (pneumatic)
sprinklers, fire protection	boiler construction and repairs
insulation of mechanical work	

Electrical contracting:

electrical transmission lines	electrical signs
electrical work for buildings and structures	telephone lines, ducts
underground electrical conduit installation	cable TV
sound and intercom, fire detection, signal and burglar alarm systems	electrical temperature controls

You can pick any of these trades as a specialty except Heavy construction, Highway, railroad and airport construction, Municipal and utility construction, and Building contracting. The Board considers these as General Contracting classifications of work. To qualify for any trade you need five years of experience in it. You also have to give the Board a list of equipment you have available to use in the trade.

The Board requires you to have a minimum business-related net worth for any classification. For Heavy construction and Building contracting it's $20,000 and for Mechanical and Electrical contracting it's $15,000. The specialty trades are $5,000 each. The Board also requires contractors to post a bond for $10,000.

You'll also need to pass a two-hour exam on contractor's business and law. The exam is open book with 50 multiple choice questions. Here's a breakdown of the content of the exam:

Content area	Number of questions
Contract management	10
Project management	10
Contractor's licensing	5
Financial management	5
Personnel regulations	5
Code of Federal Regulations	4
Insurance and bonding	3
Tax laws	3
Lien laws	3
Business organization	2

Contractor's license fees: The laws and business subjects in the exam come from the Arkansas Contractor's Reference Manual. You can buy this book from the Board for $35.

It will cost you $100 nonrefundable to file an application for a license. The exam costs $35. The license is good for one year.

Arkansas doesn't accept a license in another state for this type of license.

Plumber's Licenses

You must have a license to do plumbing work in Arkansas unless you work only on your own residence or agricultural buildings. The agriculture buildings must be outside city limits and they mustn't be connected to a public water line, sewer, or gas line. To apply for a license, contact:

Arkansas Department of Health
Plumbing and Natural Gas Section
Division of Protective Health Codes
4815 West Markham Slot 24
Little Rock, AR 72205-3867
(501) 661-2642

Arkansas issues master, journeyman, and apprentice licenses. You'll have to take an exam to get a master or journeyman license. To take the master exam you need to have five years of plumbing experience or be a professional engineer in plumbing engineering. For the journeyman plumber exam you need four years of plumbing experience. If you've had a journeyman license for a year, you can take the master exam.

Plumber's license fees: The master plumbing exam costs $125, and the license costs $200. The journeyman exam costs $75, and the license $75. An apprentice license costs $25. All licenses are good for a year, but they all expire on December 31 each year.

If you have a plumbing license in another state, ask for the reciprocal licensing form. The state will review the information you give them and may decide to accept that license. You'll still have to pay $325 for a master's license or $150 for a journeyman's license.

Electrician's Licenses

You don't need a state license to do electrical work in Arkansas. However you may need a license from any municipality you want to work in. If you get a state license you'll still have to get the municipal license and pay the fee for it but you won't have to take any exam the municipality requires for its license. If you have a valid state license, you're exempt from any municipal exam. If you want to work on any public works projects in Arkansas, you'll need a state license.

You can get a master, journeyman, or industrial maintenance electrician's license from the state of Arkansas. To apply, contact:

Contact

Arkansas Board of Electrical Examiners
10421 West Markham
Little Rock, AR 72205
(501) 682-4549
Fax: (501) 682-1765

To qualify for a master electrician exam you need:

- Electrical Engineering degree and one year experience, or
- Six years experience including two years as journeyman, or
- training and experience equivalent approved by the Board

To qualify for a journeyman electrician exam you need:

- Four years experience, or
- training and experience equivalent approved by the Board

To qualify for an industrial maintenance exam you need:

- Four years experience, or
- training and experience equivalent approved by the Board

Fee

Electrician's exam fees: The exam costs $60 for master or journeyman, and $25 for industrial maintenance electrician.

Fee

Electrician's license fees: A master license will cost you $50 and it's $25 for a journeyman or industrial maintenance license. All licenses are good for one year. Renewal costs $50 per year for master electrician, and $25 per year for journeyman and industrial maintenance electricians.

If you're already licensed in another state, Arkansas may accept your license. You'll need to give them proof of your valid license and pay the fees they charge for the Arkansas license. Check with the Board on this.

Asbestos Abatement License and Certificates

To do asbestos removal in Arkansas you need to be licensed by the Arkansas Department of Pollution Control and Ecology. To work in abatement of asbestos materials you need to get a certificate from the Department. To apply for the license and/or certificate, contact:

Contact

Pollution Control and Ecology
Attn: Asbestos Section
P.O. Box 8913
Little Rock, AR 72219-8913
(501) 682-0744

The Department issues four types of certificates:

- consultant supervisor
- contractor supervisor
- air technician
- worker

You need at least one year of experience to apply for a contractor's license. You'll have to pass a written exam with 100 multiple choice questions for a contractor's license, and one with 50 multiple choice questions for a worker's license. The Department also requires a training course — 32 hours for contractors and 24 hours for workers. You also need at least $500,000 of liability coverage.

A license or certificate is good for one year. You need to renew it by December 31 each year.

Boiler Installation License

If you sell or install boilers, pressure vessels, or pressure piping in Arkansas, you must have an Installation License. To get one, contact:

Contact

Boiler Inspection Division
Arkansas Department of Labor
10421 West Markham, Room 310
Little Rock, AR 72205
(501) 682-4513
Fax: (501) 682-4562

Boiler installation license fees: There's no exam for this license. It costs $75. The license is good for a calendar year, and you have to renew it before January 31 each year.

Alarm Systems Installation License

To install alarms systems you need to be licensed by the state. To get an application for a license, contact:

Arkansas State Police
Special Services Section
#1 State Police Plaza Drive
Little Rock, AR 72209
(501) 618-8600

The state issues two types of licenses based on how many employees you have in your company. For more than five employees, the license is called a Class E license. For five or less, it's called a Class F license. For either license, you need to pass an exam on Arkansas rules and regulations on alarm systems.

Arkansas requires all licensees to have a public liability insurance policy for at least $10,000. If your alarm system business issues Underwriters' Laboratories certificates, you'll need a $300,000 public liability insurance policy.

Alarm Systems Installation license fees: A Class E license costs $450 and a class F costs $225. Either license is good for one year.

Department of Transportation (DOT)

You must be prequalified by the Arkansas State Highway Commission to bid on projects as a prime contractor. If you're licensed in Arkansas, you can send a copy of your Arkansas Contractor's License instead of the Prequalification Questionnaire. There are a few extra requirements with this option, so contact the Commission for details before you use it.

To get a Prequalification Questionnaire, contact:

Arkansas State Highway and Transportation Department
Programs and Contacts Division
P.O. Box 2261
Little Rock, AR 72203
(501) 569-2261

The Prequalification Questionnaire will ask you for:

- your financial statement (prepared by a CPA)
- a detailed list of your equipment
- your experience record

The prequalification is good for one year, with a four-month grace period.

Essentially, you must be licensed by the Arkansas State Contractors Licensing Board before you can bid with the Department of Transportation. You can bid on a federal-aid highway project without it, but if you're awarded the project, you'll need the license to actually get the job.

Out-of-State Corporations

Any incorporated business must be registered with the Arkansas Secretary of State. To register, contact them at:

Contact

Secretary of State
State Capitol Room 058
Little Rock, AR 72201
(501) 682-3506

California

With a few exceptions, all businesses or individuals who work on any building, highway, road, parking facility, railroad, excavation, or other structure in California must be licensed by the California Contractors State License Board (CSLB) if the total cost of one or more contracts on the project is $300 or more. To apply for a license, contact:

Contact

Contractors State License Board
9835 Goethe Road
P.O. Box 26000
Sacramento, CA 95826

CSLB has an automated, toll-free number (1-800-321-2752) you can use to order an application or get other information. CSLB is also on the Internet at http://www.ca.gov/cslb/index.html.

To qualify for a license you must verify that you've had at least four years of experience in the last ten years as a journeyman, foreman, supervising employee, contractor, or owner-builder. You may be able to apply from one and one-half to three years of approved education and/or apprenticeship to this requirement. There are also special requirements for contractors who work with asbestos, contractors who remove hazardous substances, and contractors who install or remove underground storage tanks.

Unless you're applying for a joint venture license, you must have more than $2,500 worth of operating capital to apply for a new contractor's license.

CSLB issues these types of licenses to an individual, partnership, corporation, or joint venture:

Class A — General Engineering Contractor

Class B — General Building Contractor

Class C — Specialty Contractor

Class C has these subclassifications:

boiler, hot water heating and steam fitting	masonry
building moving and demolition	metal roofing
cabinet and mill work	ornamental metals
carpentry	parking and highway improvement
concrete	painting and decorating
drywall	pipeline
earthwork and paving	plastering
electrical (general)	plumbing
electrical signs	refrigeration
elevator installation	roofing
fencing	sanitation system
fire protection	sheet metal
flooring and floor covering	solar
general manufactured housing	steel, reinforcing
glazing	steel, structural
insulation and acoustical	swimming pool
landscaping	tile (ceramic & mosaic)
lathing	warm-air heating, ventilating
limited specialty	and air conditioning
lock and security equipment	welding
low voltage systems	well-drilling (water)

Unless you qualify for a waiver, you'll have to take the Board exam to get a license. The Board will review your application and let you know whether you need to take the exam or not.

The exam is closed book and has two parts which each take about two and one-half hours to finish. One part has about 100 multiple choice questions on law and business. Here's a summary of its content:

Section	Percent of exam
Project/job management	20
Licensing	15
Bookkeeping	15
Bid procedures	13
Safety	12
Contracts	10
Liens and dispute resolution	5
Employee issues	4
Insurance	3
Special circumstances	3

The other part of the exam will be on whatever specific trade you apply for. The CSLB will send you a study guide for the trade. It'll tell you the topics on the exam, the weight of each topic, and other materials you should study before taking the exam.

Contractor's license fee: It will cost you $250 nonrefundable to apply for a license and $50 for each trade you apply for after the first. The initial license costs $150. However you can't add a trade that requires an exam until you get your first license. A license is good for two years.

The Board requires a bond before it will give you a license. After you pass the exams they'll notify you how much the bond must be for. This will depend on the type of license you get and which trade it's in. Usually it's $7,500 but for a swimming pool trade license it's $10,000.

The Board may accept your trade experience and/or license in another state. However you'll still have to apply for a license, pass the law and business part of the Board exam, and pay the appropriate fees.

Recommended Reading for Contractor's License

📖 The basic study book for the law and business exam is *California Contractors License Law and Reference Book* which you can buy for about $20. The easiest way to get this book is to order it from:

General Services, Office of Procurement
4675 Watt Avenue
P.O. Box 1015
North Highlands, CA 95660
(916) 574-2200

Other publications CSLB recommends are:

📖 *Construction Management Guide* — Basic Business & Project Management for Contractors, National Association of State Contractors' Licensing Agencies, Box 30478, Phoenix, AZ 85046

📖 *California Employer's Tax Guide, California Franchise Tax Board*

📖 *Circular E, Employer's Tax Guide,* Internal Revenue Service

📖 *Employment Taxes and Information Returns,* Internal Revenue Service

📖 *How to Succeed with Your Own Construction Business,* Diller, S. & Diller, J., Craftsman Book Company, 6058 Corte del Cedro, P.O. Box 6500, Carlsbad, CA 92018

📖 *Bookkeeping for Builders,* Thomsett, M. C., Craftsman Book Company, 6058 Corte del Cedro, P.O. Box 6500, Carlsbad, CA 92018

📖 *Accounting and Financial Management for Builders,* National Association of Home Builders, 15th and M Streets NW, Washington, DC 20005

Recommended Reading for Contractor's License (continued)

Recommended publications on safety are:

📖 *California Contractors License Law and Reference Book* (1994)

📖 *Guide to Cal/OSHA*

📖 *Cal/OSHA Guide for the Construction Industry*

📖 *Guide to Developing Your Workplace Injury & Illness Prevention Program*

You can get these books from:

Department of Industrial Relations
P.O. Box 420603
San Francisco, CA 94142-0603
(415) 703-5281

Department of Transportation (DOT)

You don't need to be prequalified to bid on Department of Transportation (Caltrans) projects in California. However you do need to be licensed through the Contractors State License Board in the classes of work required to perform the work specified in any contract you bid on. The Department puts out a weekly listing of construction projects called "Advertisement for Bids." To get your name included on the mailing list for this publication, send a request to:

Contact

Department of Transportation
Office Engineer/Engineering Service Center
1727 30th Street, MS #43
Sacramento, CA 95816
(916) 227-6287

Out-of-State Contractors

Out-of-state contractors must register with the California Secretary of State. For information, contact:

Contact

Corporate Filing Support
Secretary of State
1500 11th Street, 3rd Floor
Sacramento, CA 95814
(916) 657-5448

Colorado

General construction contractors don't need licenses in Colorado. You will need a license to do electrical or plumbing work in the state, however.

Electrician's Licenses

Colorado issues master, journeyman, and residential wireman licenses. To apply for a license, contact:

Contact

Colorado State Electrical Board
1580 Logan Street, Suite 550
Denver, CO 80203-1941
(303) 894-2300

The Board will review your application and if you're eligible they'll send you information on the exam they require. For a master license you need to document five years of work experience in wiring residential and commercial buildings, including one year of planning and layout experience, or have an electrical engineering degree and one year of construction wiring experience. For a journeyman license you need four years of work experience with at least two years in commercial and/or industrial electrical work. For the residential wireman license you need two years of practical experience wiring one- to four-unit family dwellings.

The master exam is on:

- the *National Electrical Code*
- cost estimating for electrical installations
- procurement and handling of materials needed for electrical installations and repair
- reading electrical blueprints
- drafting and laying out electrical circuits
- practical electrical theory

The journeyman exam is on:

- Ohm's Law
- the *National Electrical Code*
- laying out and installing electrical circuits

The residential wireman exam is on residential wiring per the *National Electrical Code*.

Electrician's license fees: There's a fee of $30 for any application/license. Licenses are good for one year.

If you can prove you have a valid license in a state with qualifications at least equal to Colorado's and you've had the license for at least six months, you may be eligible for a Colorado license by endorsement. You still need to pay the fee, though.

To register as an electrical contractor in Colorado you need to get an Application for Electrical Contractor's Registration from the Electrical Board. You'll be asked about your workers' compensation and unemployment insurance coverage. You'll also have to sign an Acknowledgment of Responsibility form to register. It will cost you $100 to register.

Plumber's Licenses

To do plumbing work in Colorado you need a license. Colorado issues master, journeyman, and residential licenses. To apply for a license, contact:

Examining Board of Plumbers
1580 Logan Street, Suite 550
Denver, CO 80203-1941
(303) 894-2319

To apply for a master license you need to document five years of plumbing work experience; four years for a journeyman license; and two years for a residential license. You can apply every six hours of plumbing training in an accredited school for one hour of work experience, up to one year. You can also apply plumbing work in the military at the rate of one month for every six months, up to one year.

You'll need to pass an exam with both a written and a practical part to get your license. The written part is based on the *1994 Uniform Plumbing Code*. It covers:

- calculations and blueprint reading
- cross connections
- combination waste and vent systems
- gas piping
- hangers, supports and system protection
- indirect and special waste piping
- installation techniques

- plumbing fixtures
- sanitary drainage systems
- storm drainage systems
- traps, interceptors, separators
- vents/venting
- water distribution systems

The practical part of the exam will test how well you build a project specified by the Board. If you already hold a Colorado journeyman license, you don't have to take the practical part of the exam.

Fee

Plumber's license fees: It will cost you $45 for the exam. A master license costs $101, a journeyman license $61, and a residential license $51. All are good for two years.

If you can prove you have a valid license in a state with qualifications at least equal to Colorado's and you've had the license for at least six months, you may be eligible for a Colorado license by endorsement. You must have passed an exam on the *1994 Uniform Plumbing Code.* You'll still need to pay the license fee.

Department of Transportation (DOT)

To bid on a public project in Colorado you must be prequalified by the Colorado Department of Transportation. To get an application, contact:

Contact

Staff Construction
Prequalification Unit
4210 E. Arkansas, Room 287
Denver, CO 80222

Some of the details the Department will ask you about are your organization, personnel, financial condition, equipment, and experience. You'll also be asked what type of work your company does with your own equipment. Here are the types of work the Department considers:

Construction type:

general construction	seal coat
grading (general)	portland cement concrete
light grading	structures (general)
aggregates	small bridges
paving (general)	minor structures
bituminous concrete	

Incidental:

curb, gutter, flatwork
fencing
guardrail

landscaping
pavement marking
construction traffic control

Specialty:

pavement repair
structure repair
electrical, signals
building construction

water line
sprinkler system
other

You have to file a new application every year, or more often if the Department asks you to. You also need to file again if there's a significant change in the structure of your company.

Out-of-State Corporations

Out-of-state corporations must get a Certificate of Authority to do business in Colorado from the Colorado Secretary of State. To apply for this certificate, contact:

Division of Commercial Recordings
Department of State
1560 Broadway, Suite 200
Denver, CO 80202
(303) 894-2251

Connecticut

To work as a major contractor in Connecticut, you must register. To get an application for registration, contact:

Contact

State of Connecticut
Department of Consumer Protection
Real Estate & Professional Trades Division
165 Capitol Avenue, Room 110
Hartford, CT 06106
(860) 566-3290
Fax: (860) 566-7630

The application will ask you for:

- credit references
- evidence of insurance
- certificate of good standing in Connecticut or the state you're incorporated in and Connecticut
- three references on your knowledge and skills
- a statement that you're familiar with statues and regulations of the Division
- a trade name certificate where your business is (may not be required)
- current projects and ones completed in the last five years

Fee

Registration will cost you $500. It will expire on July 1 of the year after you get it.

Home Improvement Contractor's Certificate

If you'll be working in the home improvement field, you must get a certificate of registration with the state. To get an application for the certificate, contact:

Contact

Department of Consumer Protection
Central Licensing Unit
165 Capitol Avenue, Room 147
Hartford, CT 06106
(860) 566-3290

The certificate will cost you $160, or less, depending on how late in the year you get it. Certificates expire the last day of November each year. You'll also have to pay $100 every year to the Home Improvement Guaranty Fund and you may be required to post a bond.

Electrician's Licenses

To do electrical work in Connecticut you need a license. To get an application, contact:

Contact

Department of Consumer Protection
Central Licensing Unit
165 Capitol Avenue, Room 147
Hartford, CT 06106
(860) 566-3290

The Department issues several types of licenses which all require you to pass an exam. Generally you need to have two years of experience as a Connecticut licensed journeyman or Department-approved equivalent to apply for a contractor exam. To take the journeyman exam you need to complete an apprenticeship program or Department-approved equivalent. Here are the types of electrical licenses the Department issues:

Unlimited contractor	Limited low voltage journeyman
Unlimited journeyman	Limited telephone interconnect contractor
Limited contractor	Limited telephone interconnect journeyman
Limited journeyman	Limited line construction contractor
Limited low voltage contractor	Limited line construction journeyman

The electrical exams are open book, using the *National Electrical Code*. All the journeyman exam questions test your trade knowledge. To get a contractor license you must pass a trade exam and a business and law exam. The business and law exam is open book on company finances and state employment and labor laws in the *Business and Law Manual*. The exams are given by National Assessment Institute (NAI). You can contact them at:

Contact

National Assessment Institute
2 Mount Royal Avenue, Suite 250
Marlborough, MA 01752
(508) 624-0826

Electrician's license fees: A contractor exam will cost you $130 (this includes the *Business and Law Manual*) and a journeyman exam is $50. It'll cost you $75 nonrefundable to file an application for a contractor license. It's $45 nonrefundable for the journeyman application.

Licenses expire September 30 each year.

Connecticut currently has no reciprocity agreements with other states for electrical licenses.

Recommended Reading for the Electrical Exams

📖 *NFPA 70 - National Electrical Code,* 1996 edition, National Fire Protection Association, 1 Batterymarch Park, P.O. Box 9101, Quincy, MA 02269

📖 *Understanding and Servicing Alarm Systems,* 1990, H. William Trimmer, Butterworth Publishers

📖 *Direct Current Fundamentals,* 1986, Loper and Tedsen, Delmar Publishers, P.O. Box 6904, Florence, KY 41022

📖 *Alternating Current Fundamentals,* 1986, Duff and Herman, Delmar Publishers, P.O. Box 6904, Florence, KY 41022

📖 *Electronic Intrusion Alarms,* 3rd edition, J. E. Cunningham, Howard W. Sams & Co.

📖 *American Electricians Handbook,* 1996, 13th edition, Croft/Summers, McGraw-Hill Inc., Box 543, Blacklick, OH 43004-0543

📖 *NEMA Training Manual on Fire Alarm Systems,* National Electrical Manufacturers Association

📖 *Electric Motor Controls,* Walter N. Alerich, Delmar Publishers, P.O. Box 6904, Florence, KY 41022

📖 *National Electrical Safety Code,* ANSI C2

📖 *Telecommunication Wiring,* C. N. Herrick, C. L. McKim

📖 *Code of Federal Regulations - Title 29,* Part 1926 (OSHA), July 1995, Superintendent of Documents, U.S. Government Printing Office, Washington, DC 20402

📖 *Cableman's and Lineman's Handbook,* Kurtz and Shoemaker, McGraw-Hill Company, Blue Ridge Summit, PA 17294

📖 *Fire Alarm Signaling Systems Handbook,* Bukowski and O'Laughlin, National Fire Protection Association, 1 Batterymarch Park, P.O. Box 9101, Quincy, MA 02269

Elevator Installation, Repair and Maintenance Licenses

To do elevator installation, repair or maintenance in Connecticut you need a license. To get an application, contact:

Contact

Department of Consumer Protection
Central Licensing Unit
165 Capitol Avenue, Room 147
Hartford, CT 06106
(860) 566-3290

The Department issues four types of licenses which all require you to pass an exam. Generally you need to have two years of experience as a Connecticut licensed journeyman or Department-approved equivalent to apply for a contractor exam. To take the journeyman exam you need to complete an apprenticeship program or Department-approved equivalent. Here are the types of elevator licenses the Department issues:

Unlimited contractor	Accessibility contractor
Unlimited journeyman	Accessibility journeyman

The elevator exams are given by National Assessment Institute (NAI). You can contact them at:

National Assessment Institute
2 Mount Royal Avenue, Suite 250
Marlborough, MA 01752
(508) 624-0826

To get a contractor license you must pass a trade exam and a business and law exam. The business and law exam is open book on company finances and state employment and labor laws in the *Business and Law Manual.*

All the journeyman exam questions test your trade knowledge.

The exams use ANSI A17.1 Part XX, XX.1, A17.1, A17.3, and the *National Electrical Code* as references. The accessibility licenses also use ANSI A117.1 *Handicap Code.*

Elevator installation, repair and maintenance license fees: A contractor exam will cost you $130 (this includes the *Business and Law Manual*) and a journeyman exam is $50. It will cost you $75 nonrefundable to file an application for a contractor license. It's $45 nonrefundable for the journeyman application. Licenses expire August 31 each year.

Connecticut currently has no reciprocity agreements with other states for elevator licenses.

Fire Protection Sprinkler System Licenses

To do fire sprinkler system installation, repair, or maintenance in Connecticut you need a license. To get an application, contact:

Department of Consumer Protection
Central Licensing Unit
165 Capitol Avenue, Room 147
Hartford, CT 06106
(860) 566-3290

The Department issues unlimited and limited contractor and journeyman licenses. You need to pass an exam for any license. Generally you need to have two years of experience as a Connecticut licensed journeyman or Department-approved equivalent to apply for a contractor exam. To take the journeyman exam you need to complete an apprenticeship program or Department-approved equivalent.

The fire sprinkler system exams are given by National Assessment Institute (NAI). You can contact them at:

National Assessment Institute
2 Mount Royal Avenue, Suite 250
Marlborough, MA 01752
(508) 624-0826

To get a contractor license you must pass a trade exam and a business and law exam. The business and law exam is open book on company finances and state employment and labor laws in the *Business and Law Manual*.

All the journeyman exam questions test your trade knowledge.

The exams use *National Fire Protection Association Standards* and the *National Electrical Code* as references.

Fire protection sprinkler system license fees: A contractor exam will cost you $130 (this includes the *Business and Law Manual*) and a journeyman exam is $50. It will cost you $75 nonrefundable to file an application for a contractor license. It's $45 nonrefundable for the journeyman application. Licenses expire October 31 each year.

Connecticut currently has no reciprocity agreements with other states for fire sprinkler system licenses.

Heating, Piping and Cooling Licenses

To do heating, piping and cooling work in Connecticut you need a license. To get an application, contact:

Department of Consumer Protection
Central Licensing Unit
165 Capitol Avenue, Room 147
Hartford, CT 06106
(860) 566-3290

The Department issues unlimited and limited contractor and journeyman licenses for various heating, piping and cooling trades. You need to pass an exam for any license. Generally you need to have two years of experience as a Connecticut licensed journeyman or Department-approved equivalent to apply for a contractor exam. To take the journeyman exam you need to complete an apprenticeship program or Department-approved equivalent.

The heating, piping and cooling exams are given by National Assessment Institute (NAI). You can contact them at:

Contact

National Assessment Institute
2 Mount Royal Avenue, Suite 250
Marlborough, MA 01752
(508) 624-0826

To get a contractor license you must pass a trade exam and a business and law exam. The business and law exam is open book on company finances and state employment and labor laws in the *Business and Law Manual*. All the journeyman exam questions test your trade knowledge. There may also be questions on OSHA safety and health standards on the exams.

Fee

Heating, piping and cooling license fees: A contractor exam will cost you $130 (this includes the *Business and Law Manual*) and a journeyman exam is $50. It will cost you $75 nonrefundable to file an application for a contractor license. It's $45 nonrefundable for the journeyman application. Licenses expire August 31 each year.

Connecticut currently has no reciprocity agreements with other states for heating, cooling and piping licenses.

Recommended Reading for the Heating, Piping and Cooling Exams

- *BOCA National Mechanical Code*, Building Officials and Code Administrators Int. Inc., 405 West Flossmoor Road, Country Club Hills, IL, 60478-4900

- *BOCA National Plumbing Code*, Building Officials and Code Administrators Int. Inc., 405 West Flossmoor Road, Country Club Hills, IL, 60478-4900

- *NFPA 54 - National Fuel Gas Code*, 1996, National Fire Protection Association, 11 Tracey Ave., Avon, MA 02322

- *NFPA 58 - Standard for the Storage and Handling of Liquefied Petroleum Gasses*, National Fire Protection Association, 11 Tracey Ave., Avon, MA 02322

- *Commercial, Industrial, and Institutional Refrigeration*, W. B. Cooper

- *Electric Controls for Refrigeration and Air Conditioning*, 1988, 2nd edition, B.C. Langley, Prentice-Hall

- *Excavation & Grading Handbook*, 1991, Nicholas E. Capachi, Craftsman Book Company, 6058 Corte del Cedro, P.O. Box 6500, Carlsbad, CA 92018

- *HVAC Fundamentals*, McGraw-Hill Company, Blue Ridge Summit, PA 17294

- *HVAC Library, Vol. I, II, and III,* G. K. Hall & Co.

- *Hydraulic Institute Standards*, Hydraulic Institute

Recommended Reading for the Heating, Piping and Cooling Exams (continued)

📖 *Installation Guidelines for Solar DHW Systems in One and Two-Family Dwellings,* U.S. Department of Housing and Urban Development

📖 *Low Pressure Boilers,* 1994, F. M. Steingress, American Technical Publishers, 1155 West 175th Street, Homewood, IL 60403

📖 *Modern Refrigeration and Air Conditioning,* Althouse, Turnquist and Bracciano

📖 *Oil Burners,* Audel

📖 *Pipe Welding Procedures,* 1973, H. Rampaul, Industrial Press, Inc., 200 Madison Avenue, New York, NY 10016

📖 *Pipefitters Handbook,* 1967, 3rd edition, Forest Lindsey, Industrial Press, Inc., 200 Madison Avenue, New York, NY 10016

📖 *Plumbing Installation and Design,* American Technical Publishers

📖 *Practical Solar Energy Technology,* Prentice-Hall

📖 *Pump Selection,* Ann Arbor Science Publishers

📖 *Recommended Practices for the Installation of Underground Liquid Storage Systems,* Petroleum Equipment Institute, P.O. Box 2380, Tulsa, OK 74101

📖 *Steam Boiler Operation,* Jackson, Prentice-Hall

📖 *Modern Welding,* Althouse, Turnequist, Bowditch and Bowditch

📖 *The Plumbers Handbook,* 8th edition, Joseph P. Almond, MacMillan Publishing Co.

📖 *Trane Air Conditioning Manual,* 1965, Trane Company, 8929 Western Way, Suite #1, Jacksonville, FL 32256

📖 *Water Well Handbook,* Missouri Water Well and Pump Contractors Association

📖 *Welding Fundamentals and Procedures,* Gaylen, John Wiley & Sons, Inc.

📖 *Control Systems for Heating, Ventilating and Air Conditioning,* Van Nostrand Reinhold

📖 *HVAC Systems — Testing, Adjusting and Balancing,* SMACNA

📖 *Air Conditioning Principles and Systems: An Energy Approach,* John Wiley & Sons, Inc.

📖 *Water Processing for Home, Farm and Business,* Water Quality Association

📖 *Turf Irrigation Manual,* Telso Industries

Plumbing and Piping Licenses

To do plumbing and piping work in Connecticut you need a license. To get an application, contact:

Contact

Department of Consumer Protection
Central Licensing Unit
165 Capitol Avenue, Room 147
Hartford, CT 06106
(860) 566-3290

The Department issues several types of licenses which all require you to pass an exam. Generally you need to have two years of experience as a Connecticut licensed journeyman or Department-approved equivalent to apply for a contractor exam. To take the journeyman exam you need to complete an apprenticeship program or Department-approved equivalent. Here are the types of plumbing and piping licenses the Department issues:

Unlimited contractor

Unlimited journeyman

Limited sewer, storm and water lines contractor

Limited sewer, storm and water lines journeyman

Limited gasoline tanks and pumping equipment contractor

Limited gasoline tanks and pumping equipment journeyman

Limited water pump and condition equipment contractor

Limited water pump and condition equipment journeyman

Limited lawn sprinkler contractor

Limited lawn sprinkler journeyman

The plumbing and piping exams are given by National Assessment Institute (NAI). You can contact them at:

Contact

National Assessment Institute
2 Mount Royal Avenue, Suite 250
Marlborough, MA 01752
(508) 624-0826

To get a contractor license you must pass a trade exam and a business and law exam. The business and law exam is open book on company finances and state employment and labor laws in the *Business and Law Manual*. All the journeyman exam questions test your trade knowledge.

Plumbing and piping license fees: A contractor exam will cost you $130 (this includes the *Business and Law Manual*) and a journeyman exam is $50. It will cost you $75 nonrefundable to file an application for a contractor license. It's $45 nonrefundable for the journeyman application. Licenses expire October 31 each year.

Connecticut currently has no reciprocity agreements with other states for plumbing and piping licenses.

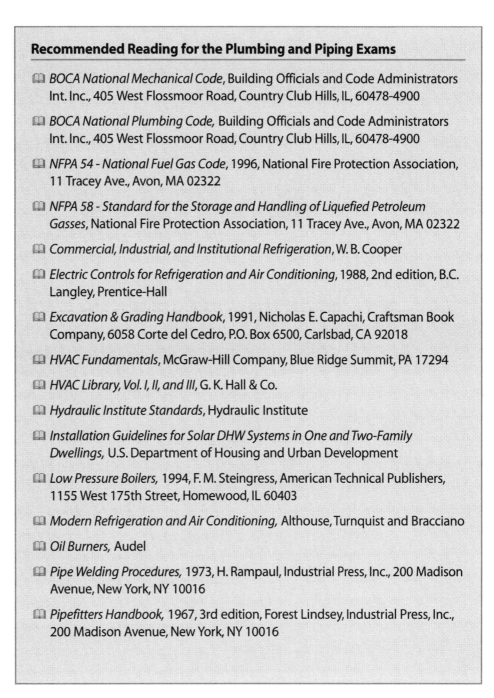

Recommended Reading for the Plumbing and Piping Exams

- *BOCA National Mechanical Code*, Building Officials and Code Administrators Int. Inc., 405 West Flossmoor Road, Country Club Hills, IL, 60478-4900

- *BOCA National Plumbing Code*, Building Officials and Code Administrators Int. Inc., 405 West Flossmoor Road, Country Club Hills, IL, 60478-4900

- *NFPA 54 - National Fuel Gas Code*, 1996, National Fire Protection Association, 11 Tracey Ave., Avon, MA 02322

- *NFPA 58 - Standard for the Storage and Handling of Liquefied Petroleum Gasses*, National Fire Protection Association, 11 Tracey Ave., Avon, MA 02322

- *Commercial, Industrial, and Institutional Refrigeration*, W. B. Cooper

- *Electric Controls for Refrigeration and Air Conditioning*, 1988, 2nd edition, B.C. Langley, Prentice-Hall

- *Excavation & Grading Handbook*, 1991, Nicholas E. Capachi, Craftsman Book Company, 6058 Corte del Cedro, P.O. Box 6500, Carlsbad, CA 92018

- *HVAC Fundamentals*, McGraw-Hill Company, Blue Ridge Summit, PA 17294

- *HVAC Library, Vol. I, II, and III*, G. K. Hall & Co.

- *Hydraulic Institute Standards*, Hydraulic Institute

- *Installation Guidelines for Solar DHW Systems in One and Two-Family Dwellings*, U.S. Department of Housing and Urban Development

- *Low Pressure Boilers*, 1994, F. M. Steingress, American Technical Publishers, 1155 West 175th Street, Homewood, IL 60403

- *Modern Refrigeration and Air Conditioning*, Althouse, Turnquist and Bracciano

- *Oil Burners*, Audel

- *Pipe Welding Procedures*, 1973, H. Rampaul, Industrial Press, Inc., 200 Madison Avenue, New York, NY 10016

- *Pipefitters Handbook*, 1967, 3rd edition, Forest Lindsey, Industrial Press, Inc., 200 Madison Avenue, New York, NY 10016

**Recommended Reading for the
Plumbing and Piping Exams (continued)**

📖 *Plumbing Installation and Design,* American Technical Publishers

📖 *Practical Solar Energy Technology*, Prentice-Hall

📖 *Pump Selection,* Ann Arbor Science Publishers

📖 *Recommend Practices for the Installation of Underground Liquid Storage Systems,* Petroleum Equipment Institute, P.O. Box 2380, Tulsa, OK 74101

📖 *Steam Boiler Operation,* Jackson, Prentice-Hall

📖 *Modern Welding,* Althouse, Turnequist, Bowditch and Bowditch

📖 *The Plumbers Handbook,* 8th edition, Joseph P. Almond, MacMillan Publishing Co.

📖 *Trane Air Conditioning Manual,* 1965, Trane Company, 8929 Western Way, Suite #1, Jacksonville, FL 32256

📖 *Water Well Handbook,* Missouri Water Well and Pump Contractors Association

📖 *Welding Fundamentals and Procedures,* Gaylen, John Wiley & Sons, Inc.

📖 *Control Systems for Heating, Ventilating and Air Conditioning,* Van Nostrand Reinhold

📖 *HVAC Systems — Testing, Adjusting and Balancing,* SMACNA

📖 *Air Conditioning Principles and Systems: An Energy Approach,* John Wiley & Sons, Inc.

📖 *Water Processing for Home, Farm and Business,* Water Quality Association

📖 *Turf Irrigation Manual,* Telso Industries

Well Drilling License

To do well drilling work in Connecticut you need a license. To get an application, contact:

Contact

Department of Consumer Protection
Central Licensing Unit
165 Capitol Avenue, Room 147
Hartford, CT 06106
(860) 566-3290

The Department issues contractor and journeyman licenses limited to well water supply drilling and contractor, and journeyman licenses limited to non-water supply drilling licenses. You need to pass an exam for any license.

To apply for either contractor exam you must prove you have at least $300,000 of liability insurance ($100,000 per person), at least $100,000 of property damage insurance ($50,000 per accident), and three years well drilling experience. To take either journeyman exam you need to complete a three-year apprenticeship program or Department-approved equivalent.

The well drilling exams are given by National Assessment Institute (NAI). You can contact them at:

Contact

National Assessment Institute
2 Mount Royal Avenue, Suite 250
Marlborough, MA 01752
(508) 624-0826

To get either contractor license you must pass a trade exam and a business and law exam. The business and law exam is open book on company finances and state employment and labor laws in the *Business and Law Manual*. For either journeyman license you only have to pass a trade exam.

The well drilling trade exams are closed book, last three hours, and have 40 multiple-choice questions. Here are the subjects on each exam and the percentage of each subject on the exam:

Well water supply drilling		Non-well water supply drilling	
Subject	**%**	**Subject**	**%**
General subject knowledge	28	General subject knowledge	28
Pumps and storage tanks	18	Pumps	20
Excavation and location	14	Excavation	14
Piping and backflow prevention	14	Piping and equipment	14
Casings and screens	12	Casings and screens	12
Yield and abandonment	12	Abandonment	12
Water quality	6		

Fee

Well drilling license fees: A contractor exam will cost you $130 (this includes the *Business and Law Manual*) and a journeyman exam is $50. It will cost you $44 nonrefundable to file an application for a contractor or well driller license. Licenses expire April 30 each year.

Connecticut currently has no reciprocity agreements with other states for well drilling licenses.

Recommended Reading for the Trade Exams

📖 *BOCA National Plumbing Code,* 1990 edition, Building Officials and Code Administrators Int. Inc., 405 West Flossmoor Road, Country Club Hills, IL, 60478-4900

Recommended Reading for the Trade Exams (continued)

◻ *Connecticut General Statutes,* Chapter 482, Section 25-128-33 through 25-128-64

◻ *Water Wells and Septic Tanks,* 2nd edition, 1992, Max & Charlotte Alth, revised S. Blackwell Duncan

◻ *Water Well Handbook,* 1984, K. E. Anderson, Missouri Water Well and Pump Contractors Association

Additional references for well water drilling exam only:

◻ *Water Processing for Home, Farm and Business,* Water Quality Association

◻ *Water Quality and Treatment,* 3rd edition, Public Water Supplies Handbook, American Water Works, Association, Inc.

You can get these books from:

Builders' Book Depot
1033 East Jefferson, Suite 500
Phoenix, AZ 85034
(800) 284-3434

Department of Transportation (DOT)

You must be prequalified to bid on Connecticut Department of Transportation work. To get an application, contact:

Contact

State of Connecticut
Department of Transportation
2800 Berlin Turnpike
P.O. Box 317546
Newington, CT 06131-7546
(860) 594-3099

Some of the details the Department will ask you about are your organization, personnel, financial condition, equipment, and experience. You'll also be asked what types of work your company wants to be prequalified in. Here are the types of work the Department uses:

Earthwork	Traffic control, illumination
Roadway subsurface, surface treatment	Railroad
Bridge structures	Incidental construction
Vertical construction	Other

You need to file for prequalification at least 30 days before bidding on a project. Prequalification is good for up to 16 months. You also need to file again if there's a significant change in the structure of your company.

You must also get a certification of registration in good standing with the Connecticut Secretary of State. See below for information on how to contact them.

Out-of-State Corporations

Any out-of-state corporation must get a Certificate of Authority for Out-of-State Corporation from the Connecticut Secretary of State. To apply for this certificate, contact:

Contact

Secretary of State for Connecticut
30 Trinity Street
Hartford, CT 06106
(860) 566-2888

Delaware

To do business in Delaware as a contractor or subcontractor, you must get a license from the Delaware Division of Revenue. You'll also need a surety bond for 6% of any contract work you do totaling $20,000 or more within a calendar year. Before you can bid on a job over $50,000 you must apply for a license. To apply for a license, ask for a Contractor's Licensing Application Packet from:

Division of Revenue
Carvel State Building
820 North French Street
Wilmington, DE 19801
(302) 577-5800
(800) 202-7826 (Delaware only)

The license will cost you $75 per year.

Asbestos Abatement Certifications

Unless you're working on your own single family dwelling, you have to be certified to handle asbestos in Delaware. However, the state has two classes of asbestos work — Class A which is for a business whose primary work is asbestos abatement, and Class B for a business that can reasonably expect some of its projects to involve asbestos abatement. Examples of Class B are roofing/siding contractors and plumbing/HVAC contractors. To apply for certification, contact:

Division of Facilities Management
O'Neill Building
P.O. Box 1401
Dover, DE 19903
(302) 739-3930

Delaware requires all asbestos supervisors and workers to have a:

- routine physical exam within year of certification
- pulmonary function test
- chest X-ray
- appropriate training course approved by Delaware
- CPR card (for supervisors only)

Fee

Asbestos abatement certification: Supervisor certification costs $50 and worker certification is $25. Both are good for one year.

If you've had asbestos training in a U.S. EPA Region III state, Delaware may accept it. These states are Maryland, Pennsylvania, Virginia, West Virginia, and the District of Columbia.

Electrician's Licenses

Delaware issues licenses for general master electricians, limited master electricians, and specialty electricians. The categories of specialty licenses are air conditioning/heating and elevators. To apply for a license, contact:

Contact

Board of Electrical Examiners
Cannon Building, Suite 203
P.O. Box 1401
Dover, DE 19903
(302) 739-4522
Fax: (302) 739-2711

Here are the requirements for master licenses:

Master general (industrial, commercial, residential)

Requirements: Four years full time work experience and two years education after high school graduation or no education and six years full time work experience

Master limited (residential)

Requirements: Two years full time practical training or equivalent technical training at a technical school or college and one year full time practical experience or no education and three years full time work experience

The Board requires you to pass an exam to get a license. It's an open book exam on the NFPA-70 1996 *National Electrical Code* book. The exam is four hours long with 80 multiple choice questions for the master electrician's general license and three hours long with 50 questions for the limited license. Here's a percentage breakdown of the subjects on the exam by license type:

Subject	Master license type	
	General	Limited
Grounding and bonding	11%	10%
Services, feeders and branch circuits	13%	18%
Raceways and enclosures	11%	8%

| | Master license type | |
Subject	General	Limited
Conductors	9%	14%
Motors and controls	11%	3%
Utilization and general use equipment	11%	15%
Special occupancies and equipment	6%	7%
General knowledge and calculations	25%	23%
Low voltage circuits	3%	2%

Exams are given by:

Contact

National Assessment Institute
261 Connecticut Drive
Burlington, NJ 08016
(609) 387-7944

After you take the exam, NAI will send your score directly to the Board. If you scored at least 75%, the Board will send you their information packet on how to proceed to get your license.

Fee

Electrican's license fees: It will cost you $10 nonrefundable to file an application for a license. The exam costs $60 nonrefundable. NAI has a practice exam they sell for $25. A license is good for two years. However the two-year period is set by the Board so your license and its cost will depend on when you get it.

Currently the Board doesn't have any reciprocity agreements with other states.

Recommended Reading for the Electrical Exam

📖 *NFPA 70 - National Electrical Code,* 1996 edition, National Fire Protection Association, 1 Batterymarch Park, P.O. Box 9101, Quincy, MA 02269

📖 *Direct Current Fundamentals,* 1986, Loper and Tedsen, Delmar Publishers, P.O. Box 6904, Florence, KY 41022

📖 *Alternating Current Fundamentals,* 1986, Duff and Herman, Delmar Publishers, P.O. Box 6904, Florence, KY 41022

📖 *Electric Motor Controls,* Walter N. Alerich, Delmar Publishers, P.O. Box 6904, Florence, KY 41022

📖 *American Electricians Handbook,* 1996, 13th edition, Croft/Summers, McGraw-Hill Inc., Box 543, Blacklick, OH 43004-0543

Recommended Reading for the Electrical Exam (continued)

📖 *Cableman's and Lineman's Handbook,* Kurtz and Shoemaker, McGraw-Hill Company, Blue Ridge Summit, PA 17294

You can get these books from:

Builders' Book Depot
1033 East Jefferson, Suite 500
Phoenix, AZ 85034
(800) 284-3434

Department of Transportation (DOT)

To bid on Delaware Department of Transportation projects you fill out a short form called Registry of Construction Contractors. To get the form, contact:

Contact

Department of Transportation
P.O. Box 778
Dover, DE 19903
(302) 739-4318
Fax: (302) 739-6119

On the form, you specify whether you want to bid on highway, bridge, and/or building construction.

Out-of-State Corporations

Corporations doing business in Delaware must register with the Delaware Secretary of State to do business in the state. For information, contact:

Contact

Division of Corporations
John G. Townsend Building
P.O. Box 898
Dover, DE 19903
(302) 739-3073
Fax: (302) 739-3812

District of Columbia

Electrical, plumbing, and home improvement contractors must be licensed to work in the District of Columbia. To get an application, contact:

Department of Consumer & Regulator Affairs
Application and Support Division
614 H Street NW, Room 407
Washington, DC 20001
(202) 727-7454

To get a license, you must pass an exam. You must also carry liability insurance and be bonded.

If you don't reside in the District of Columbia, you'll need to designate a District resident to act as your registered agent for legal purposes.

All corporations and partnerships must get a Certificate of Authority to do business in the District of Columbia from the Corporation Division of the Department of Consumer and Regulatory Affairs. To apply for this certificate, contact:

Corporation Division
Business Regulation Administration
Department of Consumer and Regulatory Affairs
614 H Street NW
Washington, DC 20001
(202) 727-72781

Florida

You need to be registered or certified to do construction work in Florida. You can get a registered contractor's license at the local level in the state but if you get a certified license you'll be allowed to work anywhere in the state. To get an application for a certified license, contact:

Contact

Business & Professional Regulation
Construction Industry Licensing Board
Attention: Examination
7960 Arlington Expressway, Suite 300
Jacksonville, FL 32211-7467
(904) 727-6530

Certified Contractor's Licenses

Here's a list of the certified contractor's licenses the Board issues. Certified licenses marked with an asterisk require one year of work experience.

General*   Commercial pool/spa*
Building*   Residential pool/spa
Residential*   Pool/spa
Mechanical   Solar
Sheet metal   Air conditioning A*
Roofing   Air conditioning B*
Plumbing   Pollutant storage
Underground utility   Specialty structure
Gypsum drywall

For some of these trades, you can qualify for a certified contractor's license by having a certified license for a specified time in another trade. Here's a list of those trades:

Building or Residential license for at least the previous four years
Trade license you qualify for: General

Residential license for at least the previous three years
Trade license you qualify for: Building

Pool servicing license for at least the previous four years or Residential pool license for at least the previous year
Trade license you qualify for: Commercial pool

Pools servicing license for at least the previous three years
Trade license you qualify for: Residential pool

Class C air conditioning license for at least the previous four years or Class B air conditioning license for at least the previous year
Trade license you qualify for: Class A air conditioning

Class C air conditioning license for at least the previous three years
Trade license you qualify for: Class B air conditioning

To qualify for the exam the Board requires for a certified license, you need to fulfill one of the following requirements:

- hold an active certified or registered Florida contractor's license
- hold a four year construction-related degree from an accredited college and one year work experience in the trade you're applying in
- hold a four year nonconstruction-related degree from an accredited college and one year work experience as a workman and one year as a foreman
- four years work experience as a workman or foreman, at least one year of the time as a foreman
- two years work experience as a workman, one year as a foreman, and one year accredited college-level courses in an appropriate field
- two years work experience as a workman, one year as a foreman, and a two year nonconstruction degree
- two years work experience as a workman, with at least one year as a foreman, and a two year construction degree applicable to the trade you're applying for

If you have a trade license from another state, you may qualify for the trade exam if you can prove you got your license by passing an exam and that your license required experience equal to what the Board requires. Other than this Florida doesn't have any reciprocity agreements other states.

The exam takes two days. It's open book, multiple choice with one part on business and finance and the other on the specific trade you want to qualify in.

You need to bring some reference books to the exam and use them to answer the test questions. You are allowed to underline in the books but you can't make any other marks or notes in them.

Once you qualify for an exam, the Board will send you a Candidate Testing Information Booklet with information on the exam.

The application and exam will cost $354 nonrefundable. The certificate costs $209 and it's good for two years.

Recommended Reading for the Contractor's Exam

- *The Guide to Florida's Construction Lien Law*, 3rd edition, N. H. Butler and L. F. Sisson, III, Profit Guard, P.O. Box 10354, Tallahassee, FL 32303

- *Florida Construction Law Manual*, L. R. Leiby, 3rd edition, 1995, Shepard's/McGraw Hill Inc., 555 Middle Creek Pkwy., P.O. Box 350, Colorado Springs, CO 80935

- *Contractors Manual*, 2nd edition with 1995 revisions, Assoc. of Builders and Contractors Institute, Inc., 4700 NW 2nd Ave., Boca Raton, FL 33431

- *Builder's Guide to Accounting*, 1987, Thomsett, Craftsman Book Company, 6058 Corte del Cedro, P.O. Box 6500, Carlsbad, CA 92018

- *Building Estimators Reference Book*, Waler's, 25th edition, F. R. Walker Co., P.O. Box 318, 1989 University Lane, Lisle, IL 60532

- *Energy Code Excerpts, A Study Guide for the 1993 Florida Energy Efficiency Code for Building Construction*, 1995, State of Florida, Department of Community Affairs, Energy Code Program, 2730 Centerview Dr., Tallahassee, FL 32399

- *AIA Document A201 - General Conditions of Contract*, 1987, American Institute of Architects, P.O. Box 60, Williston, VT 05495

- *AIA Document A401 - Contractor-Subcontractor Agreement*, 1987, American Institute of Architects, P.O. Box 60, Williston, VT 05495

- *AIA Document A701 - Instructions to Bidders*, 1987, American Institute of Architects, P.O. Box 60, Williston, VT 05495

- *AIA Document G701 - Change Order*, 1987, American Institute of Architects, P.O. Box 60, Williston, VT 05495

- *AIA Document G702 - Application and Certificate for Payment*, 1992, American Institute of Architects, P.O. Box 60, Williston, VT 05495

- *AIA Document G703 - Continuation Sheet*, 1992, American Institute of Architects, P.O. Box 60, Williston, VT 054951994

- *AIA Document G706A - Contractor's Affidavit of Release of Liens*, 1987, American Institute of Architects, P.O. Box 60, Williston, VT 05495

- *Commentary and Recommendations for Handling, Installing and Bracing Metal Plate Connected Wood Trusses*, HIB-91, Truss Plate Institute, 583 D'Onofrio Drive, Madison, WI 53719

- *Principles and Practices of Heavy Construction*, 4th edition, R. C. Smith and C. K. Andres, Prentice-Hall Publishers, 200 Old Tappan, Old Tappan, NJ 07675

- *Design and Control of Concrete Mixtures*, 1988/1990, 13th edition, Portland Cement Association, 5420 Old Orchard Road, Skokie, IL 60077-1083

Recommended Reading for the Contractor's Exam (continued)

📖 *Placing Reinforcing Bars, Recommended Practices,* 1992, 6th edition, Concrete Reinforcing Steel Institute, P.O. Box 6996, Alpharetta, GA 30239-6996

📖 *Recommended Specifications for the Application and Finishing of Gypsum Board-GA-216-93,* 1993, Gypsum Association, 810 First Street NE, Washington, DC 20002

📖 *Formwork for Concrete,* 1995, Revised, American Concrete Institute, P.O. Box 19150, Detroit, MI 48219-0150

📖 *Standard Building Code 1994,* Southern Building Code Congress International, Inc., 900 Montclair Road, Birmingham, AL 35213-1206

📖 *Blueprint Reading for the Building Trades,* 1985, Craftsman Book Company, 6058 Corte del Cedro, P.O. Box 6500, Carlsbad, CA 92018

📖 *Code of Federal Regulations - Title 29,* Part 1926 (OSHA), July 1995, Superintendent of Documents, U.S. Government Printing Office, Washington, DC 20402

Electrician's Licenses

To do electrical work statewide in Florida you need to be certified by the Electrical Contractors Licensing Board. To apply for certification, contact them at:

Contact

Electrical Contractors Licensing Board
1940 N. Monroe Street
Tallahassee, FL 32399-0071
(904) 488-3109

The Board issues these types of certificates:

Electrical contractor

Alarm contractor I (all alarm systems)

Alarm contractor II (excludes fire alarm systems)

Specialty

- residential
- lighting maintenance
- elevators
- outdoor signs
- low voltage systems
- proprietary alarms

To qualify for the exam the Board requires for a certificate, you need to fulfill one of the following requirements:

- at least three years of management experience in the trade within the last six years (half can be an approved education)

- at least four years of experience as a foreman , supervisor, or contractor in the trade within the past eight years
- at least six years of comprehensive training, technical education, or broad experience with electrical or alarm system installation or service within the past twelve years
- three years as a licensed engineer

The exam is open book on three subjects — general business, technical knowledge, and safety. Here's the information on each type of certificate and the percentage of the exam that's devoted to each subject:

Certificate type	business	% of exam on technical knowledge	safety
Electrical contractor	33	64	3
Alarm contractor I (all alarm systems)	25	66	9
Alarm contractor II (excludes fire alarm systems)	25	70	5
Residential	33	59	8
Lighting maintenance	25	70	5
Elevators	25	65	10
Outdoor signs	25	70	5
Low voltage systems	25	69	6
Proprietary alarms	6	82	12

Electrician's license fees: It will cost you $150 nonrefundable to file an application and $150 for the exam. The certificate costs $305. A certificate is good for two years.

With all these certificates, if you can prove you have a valid license in a state with qualifications at least equal to Florida's, you can get a Florida certificate without taking an exam. You still need to pay the fees though.

Department of Transportation (DOT)

To bid on work for the Florida Department of Transportation costing more than $250,000 you must be prequalified by the Department. To get an application, contact:

Contracts Administration Office
Florida Department of Transportation
Room 60, MS #55
605 Suwannee Street
Tallahassee, FL 32399-0455
(904) 414-4000

Florida: Department of Transportation

Some of the details the Department will ask you about are your organization, personnel, financial condition, equipment, and experience. They will use this information to set bid limits for your company. You'll also be asked what type of work you want to be qualified for. Here are the types of work the Department uses:

Bridges (major, intermediate, minor)

Bascule bridge rehabilitation

Grading

Drainage

Flexible paving

Portland cement concrete paving

Hot plant-mixed bituminous

Specialty work

- electrical
- fencing
- guardrails
- grassing, seeding and sodding
- landscaping
- traffic signals

- computerized traffic control systems
- bridge painting
- pavement markings
- roadway signing
- other

You have to file a new application every year, or more often if the Department asks you to. You also need to file again if there's a significant change in the structure of your company.

You need to get an Occupation License in each county in which you have a separate place of business. You get the license from the tax collector of the county.

Out-of-State Corporations

Out-of-state corporations and limited partnerships need to file an application before doing any business in the state. To get an application, contact:

Contact

Department of State
Office of the Secretary of State
409 East Gaines (Old Jail)
Tallahassee, FL 32399
(904) 488-9000

78   Contracting in All 50 States

Georgia

You don't need a license to do building contracting in Georgia, but you do need one for asbestos abatement and the mechanical trades.

Asbestos Abatement License

To do asbestos abatement work in Georgia you need a license. To get an application, contact:

Contact

Environmental Protection Division
Lead-Based Paint & Asbestos Program
4244 International Parkway, Suite 100
Atlanta, GA 30354
(404) 363-7026
Fax: (404) 362-2693

Some of the information you'll need to give on the application is:

- information on three asbestos projects you've successfully completed or supervised

- any violations you've received for asbestos work

- documentation that you've completed an approved 32 hour training course in asbestos removal

- standard operating procedures you will follow during asbestos removal

If you have a valid asbestos contractor license in another state, you may be able to get a reciprocal Georgia license. The requirements for your license must be equal to those in Georgia. The state you're licensed in must also allow a valid Georgia licensee to get a reciprocal license there.

Fee

Asbestos abatement license fee: The license will cost you $100 and it's good for three years.

Electrician's Licenses

To do electrical work in Georgia you need an electrical contractor license. To get an application for a license, contact:

Contact

State Construction Industry Licensing Board
Division of Electrical Contractors
166 Pryor Street SW
Atlanta, GA 30303-3465
(404) 656-2448
Fax: (404) 657-4220

The Division issues these types of electrical licenses:

Electrical contractor Class I — restricted to work on single phase electrical systems that aren't more than 200 amperes

Electrical contractor Class II — unrestricted

Low voltage contractor:
  alarm systems
  general systems
  telecommunications systems
  unrestricted

You must pass an exam to get a license. To qualify for the electrical contractor exam you need four years of Division-approved experience in electrical work. You can use up to two years of Division-approved education for one year of the four. For the Class II exam you need experience installing systems that are greater than single phase, 200 amperes.

The exams are given by LGR Examinations. You can contact them at:

Contact

LGR Examinations
Attn: GA CILB
1315 South Allen Street
State College, PA 16801-5992
(800) 877-3926

The exams are based on field experience and knowledge of trade practices. You can bring the recommended references to the exam.

To qualify for a low voltage electrical contractor exam, you need one year of Division-approved experience. You can use Division-approved education for up to six months of the year.

Fee

Electrician's license fees: It will cost you $30 nonrefundable to file an application for a license. An electrical contractor exam costs $90. A low voltage contractor exam costs $79. The license will cost $75 and it's good for two years.

If you have a valid license in another state, the Division may grant you a Georgia license by endorsement. Your license must have required you to pass an exam and have work experience basically equal to those for a Georgia license. Currently the Division accepts valid unrestricted electrical contractor licenses from Alabama and South Carolina. In any case, you still must pay the application and license fees for the Georgia license.

Recommended Reading for the Electrical and Low Voltage Contractor's Exam

---

Electrical Contractor's Exam:

- 📖 *American Electricians Handbook*, 1996, 13th edition, Croft/Summers, McGraw-Hill Inc., Box 543, Blacklick, OH 43004-0543

- 📖 *Code of Federal Regulations - Title 29*, Part 1926 (OSHA), July 1995, Superintendent of Documents, U.S. Government Printing Office, Washington, DC 20402

- 📖 *Employer's Tax Guide*, Circular E, Internal Revenue Service

- 📖 *NFPA 70 - National Electrical Code*, 1996 edition, National Fire Protection Association, 1 Batterymarch Park, P.O. Box 9101, Quincy, MA 02269

- 📖 *National Electrical Code Blueprint Reading*, K. L. Gebert, American Technical Publishers, 1155 West 175th Street, Homewood, IL 60403

- 📖 *Ugly's Electrical References*, 1996, G. V. Hart, United Printing Arts, 3509 Oak Forest, Houston, TX 77018

Low Voltage Contractor's Exam:

- 📖 *Code of Federal Regulations - Title 29*, Part 1926 (OSHA), July 1995, Superintendent of Documents, U.S. Government Printing Office, Washington, DC 20402

- 📖 *Employer's Tax Guide*, Circular E, Internal Revenue Service

- 📖 *NFPA 70 - National Electrical Code*, 1996 edition, National Fire Protection Association, 1 Batterymarch Park, P.O. Box 9101, Quincy, MA 02269

- 📖 *NFPA 72-1996, National Fire Alarm Code*, National Fire Protection Association, 1 Batterymarch Park, P.O. Box 9101, Quincy, MA 02269

- 📖 *NFPA 101- Life Safety Code*, 1994, National Fire Protection Association, 1 Batterymarch Park, P.O. Box 9101, Quincy, MA 02269

- 📖 *Installation and Classification of Mercantile and Bank Burglar Alarm Systems*, U.L. #681, Underwriters Laboratories, Inc.

- 📖 *Proprietary Burglar Alarm Systems*, U.L. #1076, Underwriters Laboratories, Inc.

Conditioned Air Contractor's License

To do conditioned air work in Georgia you need a conditioned air contractor license. To get an application for a license, contact:

Contact

State Construction Industry Licensing Board
Division of Conditioned Air Contractors
166 Pryor Street SW
Atlanta, GA 30303-3465
(404) 656-3939
Fax: (404) 657-4220

The Division issues two types of conditioned air licenses:

Conditioned air contractor Class I — restricted to 175,000 Btu of heating and 60,000 Btu of cooling

Conditioned air contractor Class II — unrestricted

You must pass an exam to get a license. To qualify for the conditioned air contractor exam you need three years of Division-approved experience in conditioned air work. You can use up to two years of Division-approved education for one year of the three. For the Class II exam you need experience installing systems that are greater than 175,000 Btu of heating and 60,000 Btu of cooling.

The exams are given by LGR Examinations. You can contact them at:

Contact

LGR Examinations
Attn.: GA CILB
1315 South Allen Street
State College, PA 16801-5992
(800) 877-3926

You can bring the recommended references to the exam.

Fee

Conditioned air contractor's license fees: It will cost you $30 nonrefundable to file an application for a license. An exam costs $90. A license will cost $75 and it's good for two years.

If you have a valid license in another state, the Division may grant you a Georgia license by endorsement. Your license must have required you to pass an exam and have work experience basically equal to those for a Georgia license. Currently the Division accepts some valid contractor licenses from South Carolina, North Carolina, and Texas. In any case, you still must pay the application and license fees for the Georgia license.

Recommended Reading for the Conditioned Air Contractor's Exam

NFPA 70 - National Electrical Code, 1996 edition, National Fire Protection Association, 1 Batterymarch Park, P.O. Box 9101, Quincy, MA 02269

Manual D - Duct Design Procedures for Residential Winter and Summer Air Conditioning and Equipment Selection, 1995, Air Conditioning Contractors of America, 1712 New Hampshire Ave., NW, Washington, DC 20009

Manual J - Load Calculation for Residential Winter and Summer Air Conditioning, 1986, 7th edition, Air Conditioning Contractors of America, 1712 New Hampshire Ave., NW, Washington, DC 20009

Refrigeration and Air Conditioning Technology, 1995, 3rd edition, W. Whitman and W. Johnson, Delmar ITP, P.O. Box 15015, Albany, NY 12212

Georgia Standard Gas Code (1994 Standard Gas Code, with 1995 Georgia amendments), Southern Building Code Congress International, Inc., 900 Montclair Road, Birmingham, AL 35213-1206

Georgia Standard Mechanical Code (1994 Standard Mechanical Code, with 1995 Georgia amendments), Southern Building Code Congress International, Inc., 900 Montclair Road, Birmingham, AL 35213-1206

Code of Federal Regulations - Title 29, Part 1926 (OSHA), July 1995, Superintendent of Documents, U.S. Government Printing Office, Washington, DC 20402

Employer's Tax Guide, Circular E, Internal Revenue Service

Safety Manual, 1995, Air Conditioning Contractors of America, 1712 New Hampshire Ave. NW, Washington, DC 20009

Trane Ductulator, 1976, Trane Company, 8929 Western Way, Suite #1, Jacksonville, FL 32256

Plumber's Licenses

To do plumbing work in Georgia you need a plumber's license. To get an application for a license, contact:

Contact

State Construction Industry Licensing Board
Division of Master Plumbers and Journeyman Plumbers
166 Pryor Street SW
Atlanta, GA 30303-3465
(404) 656-3939
Fax: (404) 657-4220

The Division issues three types of plumbing licenses:

Master plumber Class I — restricted to single-family dwellings and one-level dwellings designed for two families or less and commercial structures 10,000 square feet or less

Master plumber Class II — unrestricted

Journeyman plumber

You must pass an exam to get a license. To qualify for the master plumber Class I exam you need five years of Division-approved experience in plumbing work. For the Class II exam you need work experience in commercial or industrial plumbing. You can use up to two years of Division-approved education for one year of the five.

To qualify for the journeyman exam you need three years of Division-approved experience in plumbing work.

The exams are given by LGR Examinations. You can contact them at:

Contact

LGR Examinations
Attn: GA CILB
1315 South Allen Street
State College, PA 16801-5992
(800) 877-3926

You can bring the recommended references to the exam.

Fee

Plumber's license fees: It will cost you $30 nonrefundable to file an application for a license. An exam costs $90. A license will cost $75 and it's good for two years.

If you have a valid license in another state, the Division may grant you a Georgia license by endorsement. Your license must have required you to pass an exam and have work experience basically equal to those for a Georgia license. In any case, you still must pay the application and license fees for the Georgia license.

Recommended Reading for the Plumbing Exam

📖 *Blueprint Reading for Plumbers, Residential and Commercial*, 1989, D'Archangelo, D'Archangelo, and Guest, Delmar ITP, P.O. Box 15015, Albany, NY 12212

📖 *Georgia Standard Plumbing Code (1994 Standard Plumbing Code, with 1995 Georgia amendments)*, Southern Building Code Congress International, Inc., 900 Montclair Road, Birmingham, AL 35213-1206

📖 *Georgia Standard Gas Code (1994 Standard Gas Code, with 1995 Georgia amendments)*, Southern Building Code Congress International, Inc., 900 Montclair Road, Birmingham, AL 35213-1206

Recommended Reading for the Plumbing Exam (continued)

📖 *Code of Federal Regulations - Title 29, Part 1926 (OSHA)*, July 1995, Superintendent of Documents, U.S. Government Printing Office, Washington, DC 20402

📖 *Mathematics for Plumbers and Pipefitters*, 1996, D'Archangelo, D'Archangelo, and Guest, Delmar ITP, P.O. Box 15015, Albany, NY 12212

📖 *Plumbing Technology: Design and Installation*, 1994, 2nd edition, Lee Smith, Delmar Publishers, P.O. Box 6904, Florence, KY 41022

📖 *The Americans with Disabilities Act: Your Responsibilities as an Employer*, 1991, EEOC

📖 *Georgia Employer's Guide*, 1996, Summers Press, Inc., Business Publishing, P.O. Box 822068, Fort Worth, TX 76182-2068

Utility Contractor's License

To do utility contracting work in Georgia you need a utility contracting license. To get an application for a license, contact:

Contact

State Construction Industry Licensing Board
Division of Utility Contractors
166 Pryor Street SW
Atlanta, GA 30303-3465
(404) 656-2448
Fax: (404) 657-4220

The Division issues a utility contractor license and utility manager and utility foreman certificates. To qualify for a utility contractor license you have to give the Division information on the organization of your company and your personnel. You must have a licensed utility manager employed at your company. You must also have a company safety program in effect. The program must include regular scheduled safety meetings for all field personnel.

Fee

Utility contractor's license fee: It will cost you $50 nonrefundable to file an application for a utility contractor license. The license costs $75 and it's good for two years.

To qualify for a utility foreman certificate you need to have completed a Division-approved safety training course. It will cost you $30 nonrefundable to file an application for a utility foreman certificate. The certificate costs $35 and it's good for two years.

You must pass an exam to get a utility manager certificate. To qualify for the exam you need two years experience as a manager or foreman on utility systems that are at least 5 feet underground. The exam is given by LGR Examinations. You can contact them at:

LGR Examinations
Attn: GA CILB
1315 South Allen Street
State College, PA 16801-5992
(800) 877-3926

You can bring the recommended references to the exam.

Utility manager certificate fees: It will cost you $30 nonrefundable to file an application for a utility manager certificate. An exam costs $90. The certificate costs $75 and it's good for two years.

Recommended Reading for the Utility Contractor's Exam

- *Code of Federal Regulations - Title 29, Part 1926 (OSHA)*, July 1995, Superintendent of Documents, U.S. Government Printing Office, Washington, DC 20402

- *Employer's Tax Guide*, Circular E, Internal Revenue Service

- *A Guide for the Installation of Ductile Iron Pipe*, 1988, Ductile Iron Pipe Research Association, 245 Riverchase Parkway, East, Suite O, Birmingham, AL 35244

- *Placing Reinforcing Bars, Recommended Practices*, 1992, 6th edition, Concrete Reinforcing Steel Institute, P.O. Box 6996, Alpharetta, GA 30239-6996

- *Standard Specifications: Construction of Roads and Bridges*, 1993 with 1996 supplement, Department of Transportation, State of Georgia, Two Capital Square, Atlanta, GA 30334

- *The Building Estimator's Reference Book,* 1995, 25th edition, W. Spradlin, Frank Walker, Co., Publishers, P.O. Box 3180, Lisle, IL 60532

Department of Transportation (DOT)

You have to be prequalified to bid on Georgia Department of Transportation bid amounts over $500,000. To get an application, contact:

Georgia Department of Transportation
Two Capitol Square SW
Atlanta, GA 30334-1002
(404) 656-5250
Fax: (404) 656-3507

Some of the details the Department will ask you about are your organization, personnel, financial condition, equipment, and experience. You'll be asked what type of work you want to be qualified for. Here are the types of work the Department uses:

General construction

Grading, base course, drainage structures
    grading
    base courses
    curb and gutter, sidewalks, catch basin, culvert pipe
    concrete culverts

Major structures, including approaches

Paving
    portland cement
    bituminous

Specialty items
    painting bridges, other structures
    roadside development, landscaping
    grassing, seeding, erosion control
    guardrails
    fencing
    signs
    electrical
    other utilities
    erection, installation of fabricated structural materials
    dredging, hydraulic dredging, embankments
    other

Out-of-State Corporations

Out-of-state corporations must register with the Georgia Secretary of State to do business in the state. For information, contact:

Contact

Corporations
2 Martin Luther King Jr. Drive
Suite 315, West Tower
Atlanta, GA 30334
(404) 656-2817
Fax: (404) 651-9059
Internet: www.sos.state.ga.us

Hawaii

Hawaii requires general engineering, general building and specialty contractors to be licensed. To be licensed in Hawaii you need:

- proof of workers' compensation insurance
- bodily injury liability — $100,00 each person, $300,000 each occurrence
- property damage liability — $50,000 each occurrence
- place of business in Hawaii — post office box not accepted
- pass an exam on business, law, and a trade

To apply for a license, contact:

Contact

Department of Commerce and Consumer Affairs
Contractors License Board
1010 Richards Street
P.O. Box 3469
Honolulu, HI 96801

The Department issues licenses in these specialty trades:

General engineering
General building
Acoustical & insulation
Mechanical insulation
Asphalt paving
Asphalt concrete patching, sealing, striping
Boiler, hot water heating, steam fitting
Cabinet, millwork, carpentry
  remodeling & repairs
Siding application
Carpentry framing
Carpet laying
Cesspool
Drywall
Electrical
Sign contractor

Painting & decorating
Pile driving, drilling, foundation
Plastering
Plumbing
Irrigation & lawn sprinkler
  systems
Refrigeration
Reinforcing steel
Roofing
Wood shingles & shakes
Composition shingle
Sewer, sewage disposal, drain,
  pipe laying
Reconditioning, repairing pipeline
Sheet metal
Structural steel

Electronic systems
Fire and burglar alarm
Elevator
Conveyor systems
Excavating, grading, trenching
Asbestos
Fire protection
Dry chemical fire repressant systems
Flooring
Glazing & tinting
Gunite
Landscaping
Tree trimming, removal
Masonry
Cement concrete
Stone masonry
Ornamental, guardrail, fencing

Steel door
Swimming pool
Swimming pool service
Hot tub and pool
Ceramic and mosaic tile
Ventilating, air conditioning
Waterproofing
Welding
Well drilling
Pumps installation
Injection well
Solar energy systems
Solar hot water systems
Solar heating, cooling
Pole & line
Classified specialist

To qualify for the exam the Department requires for licensing, you need four years of full-time supervisory work experience as a journeyman, foreman, supervisor, or contractor within the last ten years. You can use Board-approved technical or business administration training for up to one year of work experience.

The exam is given by National Assessment Institute (NAI). You can contact them at:

Contact

National Assessment Institute
354 Uluniu Street, Suite 308
Kailua, HI 96734
(808) 261-8182

The exam has two parts — one on business and law and the other on the specific trade you want to be licensed in. The exam is closed book. The business and law part of the exam covers:

Subject	Percent of subject
Project management	19
Contact management	19
License law	10
Financial management	10
Safety regulations	10
Labor laws	6
Risk management	6
Tax laws — Employer's Circular E	6
Public works	6
Lien laws	4
Business planning and organization	4

Contractor's license fees: Each part of the exam costs $50. NAI also offers a practice exam you can buy for $25. If you also get the *Contractor's Reference Manual*, the price of the practice exam goes down to $20 plus tax, shipping, and handling.

A license is good until April 30 of each even-numbered year and costs $50.

If your business is a corporation, you must be registered with the Department. To apply for registration, contact:

Business Registration Division
Department of Commerce and Consumer Affairs
State of Hawaii
P.O. Box 40
Honolulu, HI 96810

Recommended Reading for the Contractor's Exam

The main reference for the business and law exam is the *Hawaii Reference Manual for Contractors*. You can buy it from NAI. Other references are:

- *Construction Contracting*, 5th edition, 1986, R. Clough, John Wiley & Sons, Inc.

- *Construction Methods and Management*, 2nd edition, S. W. Nunnally, 1987, Prentice-Hall, Inc.

- *Builder's Guide to Accounting*, 1987, Thomsett, Craftsman Book Company, 6058 Corte del Cedro, P.O. Box 6500, Carlsbad, CA 92018

NAI says that these books may help you study for the trade exams:

- *Alternating Current Fundamentals*, Duff and Herman, Delmar Publishers, P.O. Box 6904, Florence, KY 41022

- *Cabinetmaking and Millwork*

- *Carpentry and Building Construction*, 1993, Feirer, Hutchings & Feirer, McGraw-Hill Inc., Box 543, Blacklick, OH 43004-0543

- *Handbook for Ceramic Tile Installation*, 1995, 33rd edition, Tile Council of America, P.O. Box 326, Princeton, NJ 08542-0326

- *Circulation Systems Components for Swimming Pools, Spas and Hot Tubs*, NSF Std. 50

- *Concrete Form Construction*

- *Concrete Pipe Installation Manual*, 1995, American Concrete Pipe Association, 8300 Boone Blvd., Suite 400, Vienna, VA 22182

- *Construction Contracting*, 5th edition, 1986, R. Clough, John Wiley & Sons, Inc.

Recommended Reading for the Contractor's Exam (continued)

📖 *Design and Control of Concrete Mixtures*, 1988/1990, 13th edition, Portland Cement Association, 5420 Old Orchard Road, Skokie, IL 60077-1083

📖 *Direct Current Fundamentals*, Loper and Tedsen, Delmar Publishers, P.O. Box 6904, Florence, KY 41022

📖 *Electric Motor Control*, Walter N. Alerich, Delmar Publishers, P.O. Box 6904, Florence, KY 41022

📖 *Excavation & Grading Handbook*, 1991, Nicholas E. Capachi, Craftsman Book Company, 6058 Corte del Cedro, P.O. Box 6500, Carlsbad, CA 92018

📖 *Ground Water and Wells*

📖 *Gunite and Shotcrete Concrete*

📖 *Handling and Erecting Wood Trusses (HET-80), Bracing Wood Trusses (BWT-76)*

📖 *Landscape Operations: Management, Methods and Materials*

📖 *Mathematics for Plumbers and Pipefitters*

📖 *NFPA 90A - Installation of Air Conditioning and Ventilating Systems*, 1993, National Fire Protection Association, 1 Batterymarch Park, Box 9101, Quincy, MA 02269-9101

📖 *NFPA 91- Exhaust Systems for Air Conveying of Materials*, 1995, National Fire Protection Association, 1 Batterymarch Park, Box 9101, Quincy, MA 02269-9101

📖 *NFPA 13 - Installation of Sprinkler Systems*, 1985, National Fire Protection Association, 1 Batterymarch Park, P.O. Box 9101, Quincy, MA 02269

📖 *Painting and Decorating Craftsman's Manual*

📖 *Pipefitters Handbook*, 1967, 3rd edition, Forest Lindsey, Industrial Press, Inc., 200 Madison Avenue, New York, NY 10016

📖 *The Plumbers Handbook*, 8th edition, Joseph P. Almond, MacMillan Publishing Co.

📖 *Reciprocating Refrigeration Manual*

📖 *Recommended Specifications for the Application and Finishing of Gypsum Board-GA-216-93*, 1993, Gypsum Association, 810 First Street NE, Washington, DC 20002

📖 *Removal of Smoke and Grease Laden Vapors from Commercial Cooking Equipment*, National Fire Protection Association, 1 Batterymarch Park, Box 9101, Quincy, MA 02269-9101

📖 *Safety Code for Mechanical Refrigeration*

> ### Recommended Reading for the Contractor's Exam (continued)
>
> 📖 *Suggested Minimum Standards for Residential Swimming Pools*
>
> 📖 *Turf Irrigation Manual,* Telso Industries
>
> 📖 *Welding Fundamentals and Procedures,* Gaylen, John Wiley & Sons, Inc.
>
> You can get these books from libraries in Honolulu, Pearl City, Lihue, Hilo, and Kona. Or you can buy them from:
>
> **Builders' Book Depot**
> 1033 East Jefferson, Suite 500
> Phoenix, AZ 85034
> (800) 284-3434

Department of Transportation (DOT)

To work on state projects in Hawaii you need to be registered to do business in Hawaii and have an appropriate contractor's license. You won't need these to bid a job, but you will need them to land the contract. You may also need to fill out a prequalification form from the state Department of Transportation. Usually they require this the first time you bid as a prime contractor on a department project. They'll ask you about your work experience, financial condition, and what equipment you own.

To get a Standard Qualification Questionnaire for Prospective Bidders on Public Work Contracts, contact:

Contact

Department of Transportation
869 Punchbowl Street
Honolulu, HI 96813

This prequalification is good for at least one year.

Out-of-State Corporations

Out-of-state corporations must get a Certificate of Authority to do business in Hawaii from the Hawaii Secretary of State. To apply for this certificate, contact:

Department of Commerce & Consumer Affairs
Business Registration Division
P.O. Box 40
Honolulu, HI 96810
(808) 586-2727
Fax: (808) 586 2733

Idaho

The state of Idaho doesn't license general contractors working on private sector residential or commercial projects. That's done at the local level. However the state does license plumbers, electricians, well drillers, fire protection sprinkler contractors, and public works contractors.

Electrician's Licenses

Electrical work in Idaho requires a license. Idaho issues these licenses: contractor, journeyman, master journeyman, and specialty. To apply for a license, contact:

Contact

Division of Building Safety
Electrical Bureau, Licensing Section
277 North 6th, Suite 101
P.O. Box 83720
Boise, ID 83720-0028
(208) 334-2183
Fax: (208) 334-4891

The licenses have different requirements. Let's look first at the contractor's license. To qualify for a contractor's license you need to have two years of work experience (2,000 hours per year) as a licensed journeyman electrician or employ such a person at your company. The Bureau will review your application and if you're eligible they'll send you information on the exam they require. The exam is open book with 50 multiple choice questions. It lasts two hours. Here are the exam subjects:

Subject	Percent of exam
Project management	16
Contract management	16
Licensing	14
Financial management	10
Safety	10
Labor laws	8
Risk management	8

Subject	Percent of exam
Tax laws	8
Lien law	6
Business organization	4

The exam is given by National Assessment Institute (NAI). NAI has a practice exam you can buy for $25. You can contact them at:

National Assessment Institute
200 South Cole Road
Franklin Business Park
Building 5, Suite 200
Boise, ID 83709

Electrician's license fees: It will cost you $15 to file the application for a contractor's license, $75 for the exam, and $125 to get the license. It's good until July 1 of the year after you got it.

Now let's look at the journeyman's and master journeyman's license. To qualify for a journeyman's license you need to have four years of work experience in the trade or four years as a registered apprentice electrician. For the master journeyman's license you need two years experience as a journeyman. Then the process is the same as for the contractor's license. However the journeyman's exam is three hours long and has 70 questions. The master journeyman's exam is four hours long and has 80 questions. Here's the content of each exam:

	Number of questions on exam	
Subject	Journeyman	Master journeyman
General knowledge of the electrical trade and calculations	12	19
Raceways and enclosures	10	8
Services, feeders, and branch circuits	8	9
Grounding and bonding	8	9
Conductors	8	7
Utilization and general use equipment	8	8
Special occupancies/equipment	7	8
Motor and controls	6	8
Low voltage circuits including alarms and communications	3	4

Master and journeyman exam and license fees: NAI also gives these exams and they sell a practice exam for the journeyman exam for $25.

It will cost you $15 to file an application for the journeyman or master journeyman license and $60 for either exam. The master journeyman license costs $35 and $25 for the journeyman license. Either license is good until July 1 of the year after you got it.

Idaho has a reciprocity agreement with these states for the journeyman license:

Alaska	Oregon
Montana	Utah
North Dakota	Washington
South Dakota	Wyoming

Idaho also licenses these specialty electrical trades:

Elevator, dumbwaiter, escalator, or moving walk electrical	Mobile home dealer electrical
Signs electrical	Limited energy electrical
Manufacturing or assembling equipment electrical	Irrigation sprinkler electrical
	Domestic pump installer electrical

You need two years of experience in the trade to qualify for the exam. The exam has 30 to 50 questions and lasts two to two and one-half hours depending on the trade.

Fees and duration for specialty licenses are the same as those for electrical contractor and journeyman respectively.

With all these licenses, if you can prove you have a valid license in a state with qualifications at least equal to Idaho's, and you have the work experience Idaho requires, you can get an Idaho license without taking an exam. You still need to pay the fees, though.

Recommended Reading for the Electrical Contractor's Exam

📖 *Idaho Contractor's Reference Manual,* or

📖 *Idaho Code and Rules of the Electrical Division*

📖 *Code of Federal Regulations - Title 29, Part 1926 (OSHA),* July 1995, Superintendent of Documents, U.S. Government Printing Office, Washington, DC 20402

📖 *Handy Reference Guide to the Fair Labor Standards Act*

📖 *Guide to the Idaho Labor Laws*

📖 *Idaho Workers Compensation Employer Handbook*

📖 *ADA Fact Sheet*

📖 *The ADA - Your Responsibilities as an Employer*

📖 *Employment Eligibility Verification INS Form I-9*

📖 *Circular E, Employer's Tax Guide*

📖 *A Guide for Idaho Employers*

📖 *Idaho Mechanics' Lien Law*

Recommended Reading for the
Electrical Contractor's Exam (continued)

Recommended reading for the master and journeyman exams:

📖 *National Electrical Code,* 1996 edition

📖 *American Electricians' Handbook,* 12th edition, Croft, Watt, and Summers, McGraw-Hill Company

📖 *Electric Motor Control,* Walter N. Alerich, Delmar Publishers, P.O. Box 6904, Florence, KY 41022

📖 *Alternating Current Fundamentals,* Duff and Herman, Delmar Publishers, P.O. Box 6904, Florence, KY 41022

📖 *Direct Current Fundamentals,* Loper and Tedsen, Delmar Publishers, P.O. Box 6904, Florence, KY 41022

📖 *National Electrical Code Blueprint Reading,* K. L. Gebert, American Technical Publishers, Inc.

📖 *Guide to the National Electrical Code,* Harman and Allen, Prentice Hall, Inc.

📖 *Ferm's Fast Finder Index,* Olaf G. Ferm, Ferms Finder Index Company

📖 *Idaho Code Title 54,* Chapter 10, Department of Labor and Industrial Services

📖 *Industrial Motor Control Fundamentals,* Herman and Alerich, Delmar Publishers, P.O. Box 6904, Florence, KY 41022

You can get these books from:

Builders' Book Depot
1033 Jefferson, Suite 500
Phoenix, AZ 85034
(800) 284-3434

Recommended Reading for Electrical Specialty Exam

📖 *Sign Electricians' Workbook,* 1993, James G. Stallcup, American Technical Publishers, Inc.

📖 *Neon Techniques & Handling,* Samuel C. Miller, Signs of the Times Publishing Company

📖 *Understanding Electronics,* 1989, Warring and Sloan, TAB Books, Inc.

📖 *Design and Application of Security/Fire Alarm Systems*, 1990, John Traister, McGraw-Hill Company

📖 *NEMA Training Manual on Fire Alarm Systems*, 1992, National Electrical Manufacturers Association

> **Recommended Reading for the**
> **Electrical Contractor's Exam (continued)**
>
> **Recommended Reading for Electrical Specialty Exam (continued)**
>
> 📖 *NFPA 72-1993, National Fire Alarm Code*, National Fire Protection Association, 1 Batterymarch Park, P.O. Box 9101, Quincy, MA 02269
>
> 📖 *Proprietary Burglar Alarm Systems, UL #1076*, Underwriters Laboratories, Inc.
>
> 📖 *Understanding and Servicing Alarm Systems*, 1990, H. William Trimmer, Butterworth Publishers

Plumber's Licenses

To do most plumbing work in Idaho you need a license. The state licenses plumbing contractors, journeymen, specialty contractors, and specialty journeyman. There are two types of specialty plumbing licenses: mobile home hook-up contractor and mobile home hook-up journeyman. If you hold a contractor or journeyman license, you can do specialty work without getting a specialty license. To apply for a license, contact:

Contact

Plumbing Bureau
Division of Building Safety
P.O. Box 83720
Boise, ID 83720-0068
(208) 334-3442
Fax: (208) 334-3470

To qualify for a plumbing contractor license you need at least two and one-half years of work experience as a journeyman plumber in Idaho or another state. For a journeyman plumber's license you need at least four years experience as an apprentice supervised by a qualified journeyman plumber. For a specialty license you need two years of work experience in mobile home hook-ups. You'll also have to pass a written exam to get any of these licenses.

Here's a summary of the subjects covered on the contractor and journeyman exams and the number of questions for each subject:

Subject	Number of questions
Code, basic materials, applications, and methods	35 - 40
Water supply systems	25 - 30
Drainage and sewer systems	25 - 30
Special drain systems including interceptors and traps	5 - 15

Plumber's license fees: The fee for a contractor's license is $147.50 and $87.50 for a journeyman license. All plumbing licenses expire on December 31 each year.

All plumbing contractors need to post a $2,000 compliance bond. You must do this before you can take the licensing exam. The bond expires on December 31 each year.

Idaho has a reciprocal agreement with Washington, Oregon, and Montana for the journeyman plumber license only.

If you plan to work as an apprentice plumber in Idaho you need to register and pay a $5 fee.

Recommended Reading for the Plumbing Exam

- *Uniform Plumbing Code*, 1991 edition, IAPMO (International Association of Plumbing and Mechanical Officials)

- *Idaho Code Title 54, Chapter 26*, Department of Labor and Industrial Services

- *American Water Works Association Manual*, current copy, Pacific Northwest Edition

---

Well Drilling License

Any company or individual drilling a well in Idaho must be licensed. To get an application for a license, contact:

Department of Water Resources
1301 North Orchard Street
Statehouse Mall
Boise, ID 83720-9000
(208) 327-7900
Fax: (208) 327-7866

The Idaho Water Resource Board is the agency handling well driller's licenses. Their application will ask you about your well drilling experience, what kind of well drilling rigs you use, what type of wells you intend to work on, and what references you can give. You'll also need to:

- employ at least one qualified well drilling supervisor who will be responsible for drilling work
- provide a list of all the operators you employ
- post a surety or cash bond for $5,000 to $20,000 (amount is set by the Director of the Idaho Water Resource Board)
- pay a $25 license fee

To be a qualified well drilling supervisor you must have been employed full time for at least 30 months under the supervision of someone who has a valid Idaho well driller's license (or its equivalent). You can substitute appropriate classroom work for 12 of the 30 months . You must also pass an exam given by the Director of the Idaho Water Resources Board. The exam is on:

- Idaho water law on the appropriation of ground water
- land description by section, township, etc.
- geology
- Idaho law on well drilling
- ground-water geology
- how to use well drilling equipment
- well construction standards adopted by the Board

All your operators must also be qualified by the Board before you can get a license. The board will give each one a written exam on Idaho laws on water and well drilling, land description, and well construction standards.

All well drilling licenses expire on June 30 each year.

Fire Protection Sprinkler Contractor's License

To do fire sprinkler system work in Idaho you need a Fire Protection Sprinkler Contractor's License. To qualify for the license you need to:

- be an owner, officer, or manager of your company
- have installed four fire sprinkler systems of more than 200 heads each
- have gotten Level II Certification in fire protection (Automatic Sprinkler System Design) from the National Institute for Certification in Engineering Technologies or its equivalent
- pass an exam given by the State Fire Marshal
- post a bond for at least $2,000
- have liability insurance for at least $250,000

To apply for a license, contact:

Contact

State of Idaho
Department of Insurance
700 West State Street, Third Floor
P.O. Box 83720
Boise, ID 83720-0043
(208) 334-4250
Fax: (208) 334-4398
E-Mail: lallen@doi.state.id.us

Fee

Fire protection sprinkler contractor's license fees: The license expires December 31 each year. Your first license will cost $400, and renewal is $100. The exam costs $25.

Idaho also licenses fitters who install and maintain fire sprinkler systems under the supervision of a licensed fire protection sprinkler contractor. To get this license you need to work 1,000 hours per year for three consecutive years as a fitter. Then you need to pass an exam given by the State Fire Marshal. The original license costs $50; renewal is $25. The exam costs $25.

Public Works Contractor's License

Basically, to bid or work on a public works project costing more than $5,000 in Idaho, you need a Public Works Contractor's License. To apply for a license, contact:

Contact

Public Works Contractors State License Board
1109 Main Street, Suite 480
P.O. Box 83720
Boise, ID 83720-0073
(208) 334-2966
Fax: (208) 334-2785

Idaho has six classes of licenses. Here are the classes and their fees:

Class	Estimated cost not more than	Fee
D	$50,000	$75
C	$100,000	$75
B	$250,000	$75
A	$600,000	$150
AA	$1,000,000	$150
AAA	over $1,000,000	$150

The Board will ask you for five references, information on your company structure, experience, equipment, and financial condition. You also need to specify what types of construction you want to be licensed in:

Heavy construction
Highway construction
Building construction
Specialty construction:

- acoustical - drywall
- air conditioning, heating
- blasting
- bridges & structures

- landscaping, seeding, mulching
- lath, plaster
- masonry
- guardrail, safety barriers

- building cleaning & maintenance
- chimney repair
- clearing
- communications, alarm systems
- concrete
- craning & erection
- crushing
- demolition
- drilling
- electrical
- elevators/lifts/hoist
- excavation & grading
- fencing
- fire sprinkler systems
- flooring
- floor coverings/carpeting
- glass, glazing
- hauling
- institutional equipment
- insulation
- millwork, fixtures
- ornamental metals
- painting & decorating
- paving
- plumbing
- refrigeration
- roofing & siding
- sand blasting
- sheet metal
- signs
- sprinklers/irrigation systems
- steel fabrication/erection/installation
- tile/terrazzo
- traffic marking & striping
- utilities
- waterproofing/caulking
- well drilling
- boiler, hot-water heating & steam fitting
- other

If you're a plumber or electrician you must be licensed by the Plumbing or Electrical Board in Idaho before you can get a public works license for these trades.

You'll also need to take an exam which the Board sends with the application. The exam has eight questions on license law, four on finances, and five on bonds and general information.

The license is good for one year.

Out-of-State Corporations

Out-of-state corporations must register with the Idaho Secretary of State. For information, contact:

Contact

Idaho Secretary of State
P.O. Box 83720
Boise, ID 83720-0080
(208) 334-2300
Fax: (208) 334-2847

Illinois

Most construction contractors don't need to be licensed in Illinois. The exception is the roofing contractor.

Roofing Certificate

To do roofing work in Illinois you need to be certified by the Illinois Department of Professional Regulation. To get an application, contact:

Illinois Department of Professional Regulation
320 West Washington Street, 3rd floor
Springfield, IL 62786
(217) 782-8556

You don't have to pass an exam to get a roofing certificate. But you do have to provide information about your company's organization and your workers' compensation and unemployment insurance. You'll also have to post a $5,000 bond, prove you have $100,000 of property damage insurance, and $300,000 of personal injury insurance.

Roofing certificate fees: It will cost you $50 or $100 nonrefundable to file an application, depending on when you file it. All certificates expire on June 30 each odd-numbered year.

Department of Transportation (DOT)

To bid on Illinois Department of Transportation projects you need to be prequalified. To get an application, contact:

Illinois Department of Transportation
Bureau of Construction
2300 South Dirksen Parkway
Springfield, IL 62764
(217) 782-6667
Fax: (217) 524-4922

Some of the details the Department will ask you about are your organization, personnel, financial condition, equipment, and experience. You'll be asked what type of work you want to be qualified for. Here are the types of work the Department uses:

Earthwork	Fencing
Portland cement concrete paving	Guardrail
Bituminous plant mix	Grouting
Bituminous aggregate mixtures	Painting
Miscellaneous bituminous paving	Signing
Cleaning, sealing cracks, joints	Paint pavement marking
Soil stabilization, modification	Thermoplastic pavement marking
Aggregate bases, surfaces	Epoxy pavement marking
Structures - highway, railroad, waterway	Installation of raised pavement markers
Structures repair	Pavement texturing, surface removal
Anchors, tiebacks	Cold milling, planning, rotomilling
Drainage	Erection
Drainage cleaning	Demolition
Electrical	Fabrication
Cover and seal coats	Tunnel excavation
Slurry applications	Expressway cleaning
Miscellaneous concrete construction	Railroad track construction
Landscaping	Marine construction
Seeding and sodding	Hydraulic dredging
Vegetation spraying	Hot in-place recycling
Tree trimming, selective tree removal	Cold in-place recycling

Your prequalification rating is good for 16 months from the date of the financial statement your prequalification application is based on.

Out-of-State Corporations

Out-of-state corporations must get a Certificate of Authority to do business in Illinois from the Illinois Secretary of State. To apply for this certificate, contact:

Contact

Illinois Secretary of State
Corporation Division
330 Howlett Building
Springfield, IL 62756
(217) 782-7880
Fax: (217) 782-4528

Indiana

Only plumbing contractors need to be licensed in Indiana. Public Works and Department of Transportation work must be done by certified or prequalified contractors.

Plumber's Licenses

You need a license to do plumbing work in Indiana. The state issues journeyman and contractor licenses. To get an application for a license, contact:

Contact

Indiana Professional Licensing Agency
302 West Washington Street, Room E034
Indianapolis, IN   46204-2700
(317) 232-2980

The Agency will review your application and if you're eligible they'll send you information on the exam they require. For the contractor license you need to prove you've completed one of these:

- four years in an approved apprenticeship program
- four years (6,400 hours) plumbing work experience
- four years plumbing work under the direction of a licensed plumbing contractor

For the journeyman license you need four years in an approved apprenticeship program or four years (6,400 hours) work experience in the trade.

The exam for plumbing contractor has three sections - practical, written, and drawing. The practical section has two parts - the copper pipe project and the soil pipe project. The written section of the exam is multiple choice. The drawing section is on sanitary drainage systems and water distribution systems. If you already have a journeyman's license, you only have to pass the written and drawing section to get your contractor's license.

The exam for journeyman plumber is the same as the contractor exam except it doesn't have a drawing section.

Fee

Plumber's license fees: It costs $30 to apply for the contractor exam and $15 for the journeyman exam. The license itself will cost $25 to $50 for the contractor and $10 to $15 for the journeyman depending on whether you get it in an even or odd numbered year. Either license is good for two years.

Indiana also has a Plumbers Recovery Fund which may require you to pay a nominal fee when you get your license.

If you can prove you have a valid license in a state with qualifications at least equal to Indiana's, you can get an Indiana license without taking an exam. You still need to pay the fees, though.

Department of Transportation (DOT)

Contact

To bid on Indiana Department of Transportation projects you have to be prequalified by the Department. To get an application for prequalification, contact:

Prequalification Engineer
Indiana Department of Transportation
100 N. Senate Avenue, Room N855
Indianapolis, IN 46204-2217
(317) 232-5095

Some of the details the Department will ask you about are your organization, personnel, financial condition, equipment, and experience. They may also want to have a personal interview with you. You'll be asked what type of work you want to be qualified for. Here are the five general types of work the Department uses:

Concrete base pavement and resurface
Bituminous base pavement and resurface
Grading
Bridges and approaches
Special contracts and subcontract components

Prequalification is good for one year but the Department reserves the right to ask you for another prequalification statement at any time.

Public Works Contractor's Certification

Contact

To bid on a public work project in Indiana costing more than $100,000 you must be certified by the state. To get an application, contact:

Executive Secretary
Certification Board
402 West Washington Street, Room W467
Indianapolis, IN 46204-2746
(317) 232-3005

The Board will ask you about your organization, personnel, equipment, and experience. You'll also be asked what categories of work you want to be certified for. The Board has over a hundred different categories — you can pick four to be certified in. The Board will give major consideration to your experience in a category to certify you.

Contact

Out-of-State Corporations

Any out-of-state corporation must be authorized to do business in Indiana.
For information, contact:

Secretary of State
302 W. Washington Street, Room E018
Indianapolis, IN 46204
(317) 232-6576

Iowa

Iowa requires you to register to do business in the state. You don't have to take an exam to register. To get an application, contact:

Contact

Iowa Division of Labor
1000 East Grand Avenue
Des Moines, IA 50319-0209
(515) 281-5387
Fax: (515) 281-7995

Plumbers and electricians must be registered with the state but are licensed at the local level.

Department of Transportation (DOT)

To bid on any Iowa Department of Transportation construction project, you must be prequalified by the Department. To get a Prequalification Contractor's Statement, contact:

Contact

Office of Contracts
Iowa Department of Transportation
800 Lincoln Way
Ames, IA 50010
(515) 239-1414
Fax: (515) 239-1325

The Transportation Department will ask you for these items:

- financial statement
- your current and past work projects
- what equipment your firm owns

To qualify to bid on projects costing more than $100,000 you must have a CPA fill out the contractor's statement. The Department will use this information to give your company a prequalification rating. The formula they use for your maximum prequalification amount is:

$$\frac{current\ assets - current\ liabilities + non\text{-}current\ assets - non\text{-}current\ liabilities}{2}$$

They multiply this amount times a factor from 1 to 10, based on your experience. Also you have to have completed contracts in the following total quantities to be awarded projects for more than these quantities:

Project	Quantity
Concrete paving	100,000 SY
Grading	500,000 CY
Bridges	$200,000
Culverts	$100,000
Bituminous pavements	50,000 tons

Prequalification is good for one year but the Department may ask you for another contractor's statement at any time.

You must also be registered with the Division of Labor Services. You can contact them at:

Contact

Iowa Division of Labor
1000 East Grand Avenue
Des Moines, IA 50319-0209
(515) 281-5387
Fax: (515) 281-7995

Out-of-State Corporations

An out-of-state contractor must also satisfy a special bond requirement. The Division of Labor Services also has information on this. An out-of-state corporation needs a Certificate of Authority from the Iowa Secretary of State. For information, contact:

Contact

Corporations Division
Iowa Secretary of State
Hoover Building
Des Moines, IA 50319
(515) 281-5204

Kansas

Kansas doesn't license construction contractors at the state level. But you should check for licensing requirements at the local level.

There are some registration requirements for revenue purposes in Kansas. If you're a non-resident contractor, you have to register each contract worth over $10,000 with the secretary of revenue. It will cost you $10 for each contract.

If you register with the Secretary of State (see below) you don't have to register your contracts. You do have to file a bond with the Director of Revenue in Kansas for any contract you register before you start any work on the contract. The amount of this bond is 8 percent of the contract (minimum $1,000).

To do business in Kansas, register with the Kansas Department of Revenue. You can contact them at:

Contact

Department of Revenue
State of Kansas
Docking State Office Building
915 SW Harrison St., 3rd Floor
Topeka, KS 66625-0001
(913) 296-0222
Fax: (913) 291-3614

They will send you forms and information booklets to tell you how to register contracts, get a tax number, post bonds, and so on.

Department of Transportation (DOT)

You must be prequalified to bid on Kansas Department of Transportation projects. To get the Department application for prequalification (called a Contractor's Qualification Statement and Experience Questionnaire), contact:

Contact

Kansas Department of Transportation
Docking State Office Building
Topeka, KS 66612-1568
(913) 296-3566
Fax: (913) 296-1095

Here are some of the things the questionnaire will ask:

- what construction equipment you own
- what your work experience is
- what work you've completed
- what surety company will furnish your bonds
- what classes of work you want to be prequalified in

Here are the classes of work the Department uses:

Grading	Rest area structures, buildings
All structures	Electric lighting, traffic signals
RC box structures, culverts, misc. concrete	All signing and delineation
Light surfacing	Minor signing
Base courses	Pavement marking
Plant mix bituminous mixtures	Guardrail, fencing
Portland cement concrete pavement	Bridge/structure painting
Seeding, roadside improvement	Miscellaneous

Prequalification is good for up to one year.

Out-of-State Corporations

An out-of-state corporation doesn't have to register with the Kansas Secretary of State to do business in the state. However Kansas does require any out-of-state corporation to post a bond of 8 percent (minimum $1,000) on any contract it has over $10,000. If you register with the Secretary of State, you don't need to post this bond. It will cost you $10 to register with the Secretary of State. To get an application, contact:

Contact

Secretary of State
Capital Building, 2nd Floor
Topeka, KS 66612
(913) 296-4564
Fax: (913) 296-4570

Kentucky

The only construction contractors licensed by the state of Kentucky are electrical and plumbing contractors.

Electrician's Licenses

Kentucky licenses journeyman, master, and electrical contractors. The state also certifies master electricians and electrical contractors. To get an application, contact:

Contact

Department of Housing, Buildings and Construction
Division of Fire Prevention
1047 US Highway 127 S, Suite 1
Frankfort, KY 40601-4337
(502) 564-3626
Fax: (502) 564-6799

You'll have to pass an exam to get a certificate or license. Here are the subjects on the exam and the approximate percentage of questions on each subject:

Subject	Percent of exam		
	Contractor	Journeyman	Master
Grounding and bonding	11	11	11
Services, feeders, branch circuits, overcurrent protection	13	14	11
Raceways, enclosures	11	14	11
Conductors	9	12	11
Motors, controls	11	11	9
Utilization and general use equipment	11	11	11
Special occupancies, equipment	6	6	6
General knowledge of electrical trade, calculations	25	18	25
Low voltage circuits, alarms, communications	3	3	3

Exams are given by Block & Associates. You can contact them at:

Contact

Block State Testing Services
2100 NW 53rd Avenue
Gainesville, FL 32652
(800) 280-3926

Fee

Electrician's license fees: It will cost you $100 to file an application for a certificate. An exam will cost you $50.

Kentucky doesn't have any reciprocity agreements with other states.

Recommended Reading for the Electrical Exam

- *NFPA 70 - National Electrical Code*, 1996 edition, National Fire Protection Association, 1 Batterymarch Park, P.O. Box 9101, Quincy, MA 02269

- *Alternating Current Fundamentals*, Duff and Herman, Delmar Publishers, P.O. Box 6904, Florence, KY 41022

- *American Electricians Handbook*, 1996, 13th edition, Croft/Summers, McGraw-Hill Inc., Box 543, Blacklick, OH 43004-0543

- *Direct Current Fundamentals*, Loper and Tedsen, Delmar Publishers, P.O. Box 6904, Florence, KY 41022

- *Electric Motor Control*, Walter N. Alerich, Delmar Publishers, P.O. Box 6904, Florence, KY 41022

- *Guide to the National Electrical Code*, Harman and Allen, Prentice Hall, Inc.

- *National Electrical Code Blueprint Reading*, K. L. Gebert, American Technical Publishers, 1155 West 175th Street, Homewood, IL 60403

You can get these books from:

Professional Book Sellers
2200 21st Avenue South, Suite #105
Nashville, TN 37212
(800) 572-8878

Plumber's Licenses

To operate a plumbing business in Kentucky you need a master plumber license. To install plumbing you need a journeyman license. To get an application for a plumbing license, contact:

Department of Housing, Buildings, & Construction
Division of Plumbing
1047 US Highway 127 S, Suite 1
Frankfort, KY 40601-4337
(502) 564-3580
Fax: (502) 564-6799
E-mail: TBarnes@mail.state.ky.us

Plumber's license fees: You need to be a U.S. citizen or have applied for your first papers to get a license. You also have to pass an exam. The master exam, which lasts three hours, is based on written questions and chart analysis. It will cost you $150 to take the master exam. You need to have a journeyman's license to take the master exam.

The journeyman exam has written questions, chart work, and a practical section. The chart work will ask you to complete a drawing of a layout. The written and chart sections of the exam together last two and one-half hours. The practical section lasts two and one-half hours. The journeyman exam will cost you $50.

The exams use the Kentucky State Plumbing Law, Regulation and Code book. You can get the book from the Division for $15.

Kentucky doesn't have a reciprocal agreement with any other state for plumbing its licenses.

Department of Transportation (DOT)

To bid on transportation and highway work in Kentucky you need to get a certificate of eligibility. To get an application, contact:

Kentucky Transportation Cabinet
Division of Contract Procurement
Frankfort, KY 40622
(502) 564-3500
Fax: (502) 564-8961

Some of the details you'll be asked for are your organization, personnel, financial condition, equipment, and experience. You'll be asked what type of work you want to be qualified for. Here are the types of work the Department uses:

Principal types of work:

    Grade and drain

    Signs

    Portland cement concrete paving

    Lighting

    Bituminous concrete paving

Landscaping

Bridge projects

    less than 70 ft clear span

    less than 100 ft clear span

    100 ft clear span and over

    demolition of major bridges over navigable stream

Incidental types of work:

Clearing and grubbing	Culverts
Ditching and shouldering	Bridge repair
Bridge approaches	Bridge deck repair
Guardrails	Bridge painting
Fencing	Steel erection
Seeding and sodding	Tieing steel reinforcement
Dense graded aggregate base construction	Furnish and drive piling
Dredging	
Cement concrete base construction	Hydraulic embankment construction
Soil cement base construction	
Plant mix bank gravel base construction	Storm drainage and storm sewer
Curb and gutter	Slurry seal
Sidewalk	Buildings and related construction
Entrance pavement	Demolition
Paved ditch	

The certificate of eligibility is good until 120 days after the end of your company's fiscal year.

Out-of-State Corporations

An out-of-state corporation needs a Certificate of Authority from the Kentucky Secretary of State to do business in the state. For information, contact:

Contact

Corporate Division
Secretary of State
Capitol Building
Frankfort, KY 40601
(502) 564-2848

Louisiana

To do construction work in Louisiana you need to be licensed by the State Licensing Board for Contractors. You also need to post a bond equal to 5% of the contract price for any lump sum contract or 5% of the estimated contract price for any cost — plus contract you get. This bond must be for $1,000 or more.

If you want to work on residential projects you must have a Residential Building Contractor's license to build any residence which is three stories or less in height if the cost of the project is more than $50,000.

If you want to do plumbing, asbestos abatement, or underground tank installation you will also need a license from other agencies. You'll find more information on these particular trades after the discussion of licensing by the State Licensing Board. To apply to the Board, contact:

Contact

State of Louisiana
State Licensing Board for Contractors
P.O. Box 14419
Baton Rouge, LA 70898-4419
(504) 765-2301
(800) 256-1392
Fax: (504) 765-2690

The Board issues these classifications of licenses:

Building construction
- acoustical treatments
- air conditioning, ventilation, refrigeration, duct work
- electrical construction of structures
- fire sprinklers
- foundation for building, equipment, machinery
- incinerator construction
- installation of equipment, machinery, engines
- installation of pneumatic tubes, conveyors
- insulation for cold storage and buildings

- insulation for pipes and boilers
- landscaping, grading, beautification
- lathing, plastering, stuccoing
- masonry, brick stone
- ornamental iron, structural steel erection, steel buildings
- painting, interior decorating, carpeting
- pile driving

Heavy construction

- clearing, grubbing, snagging
- dams, reservoirs, flood control work (not levees)
- dredging
- electrical transmission lines
- foundations, pile driving
- industrial piping
- industrial plants
- industrial ventilation
- oil field construction
- oil refineries
- railroads
- transmission pipeline construction
- tunnels
- wharves, docks, harbor improvements, terminals
- landscaping, grading, beautification
- fencing
- plumbing

Residential construction

- rigging, house moving, wrecking, dismantling
- roof decks
- roofing and sheet metal, siding
- sheet metal duct work
- steam and hot water heating in buildings or plants
- stone, granite, slate, resilient floor installation, carpeting
- swimming pools
- tile, terrazzo, marble
- water cooling towers and accessories
- dry walls
- driveways, parking areas, asphalt, concrete (not on highway and streets)
- fencing

Highway, street, bridge construction

- driveways, parking areas, asphalt, concrete highway, street subsurface
- permanent or paved highways, streets (asphalt hot and cold plant mix)

- permanent or paved highways, streets (asphalt surface treatment)
- permanent or paved highways, streets (concrete)
- permanent or paved highways, streets (soil cement)
- secondary roads
- undersealing, leveling roads
- earthwork, drainage, levees
- clearing, grubbing, snagging
- culverts, drainage structures
- concrete bridges, over and under passes
- steel bridges, over and under passes
- wood bridges, over and under passes
- landscaping, grading, beautification

Municipal, public works construction

- filter plants, water purification
- gas pipe work
- sewer pipe work
- storm drain pipe work
- water line pipe work
- power plants
- sewer plants, sewer disposal
- underground electrical conduit installation
- landscaping, grading, beautification
- fencing

Electrical work

- electrical transmission lines
- electrical work for structures
- underground electrical conduit installation

Mechanical work

- HVAC drainage, sewers
- industrial pipe work, insulation
- plumbing
- mechanical controls

Hazardous materials

- asbestos removal, abatement
- hazardous materials cleanup, removal
- hazardous materials site remediation

Plumbing

- potable, nonpotable water systems
- sanitary, nonsanitary waster, sewerage construction

Other

On the application the Board sends you, you'll be asked about your company's work experience and organization. You'll also have to give the Board references who are licensed in Louisiana. And you'll have to submit a financial statement with your application.

Contractor's license and exam fees: The Board requires you to pass an exam to get a license. The exam has a business and law section and a trade section. The exam costs $50 for each classification you take an exam in. You also have to pay the Board $15 for study material for the business and law part of the exam.

If you can prove you have a valid license in a trade in another state, you won't have to take the exam for that trade. You will still have to pass the business and law section though.

A license costs $100 for any one classification and $95 for each additional classification. The maximum you can pay for a license in multiple classifications is $300. A license is good for one year. All licenses expire on December 31 each year.

The Board investigates out-of-state contractors before licensing them. You must pay a $400 surcharge to cover the costs of this investigation. You must also wait 60 days from the date of your application while the Board completes the investigation.

Asbestos Abatement Certificate

In addition to the State Contractor's Board license, you'll need to be certified to do asbestos abatement work in Louisiana. To get more information on this, contact:

Office of Air Quality
Asbestos/Lead Division
P.O. Box 82135
Baton Rouge, LA 70884-2135
(504) 765-2555
Fax: (504) 765-2559

The Division issues contractor, worker, inspector, supervisor, management planner, and project designer certificates. You will need to complete appropriate Division-approved training to qualify for any certificate.

Asbestos abatement certificate fees: It will cost you $50 for a worker certificate. All other certificates cost $200.

Underground Tank Installer Certificate

To supervise underground storage tank installation, repair or closure, you must be certified by the Louisiana Department of Environmental Quality. To get an application, contact:

Louisiana Department of Environmental Quality
Underground Storage Tank Division
P.O. Box 82178
Baton Rouge, LA 70884-2178
(504) 765-0243
Fax: (504) 765-0366

The Division issues certificates for only installation/repair, only closure, or both installation/repair and closure.

You must pass an exam to get a certificate. To qualify to take an exam you must have two years of experience and have worked on at least five jobs in the type of work you are getting a license in. For the combined installation/repair/closure certificate you need experience and work on jobs in all three types of work — installation, repair and closure.

The installation/repair exam has 150 multiple choice questions and the closure exam has 100 multiple choice questions. The Division sells study guide questions and reference documents for the exams. Here's a list of the study guides and their cost:

	Study guide	Guide and references
Installation/repair	$8	$49
Closure	$8	$83
Installation/repair/closure	$12	$119

Underground tank installer's certificate exam fees: It will cost you $100 to take an exam. There's no additional cost for the certificate. Your certificate will expire on December 31 every second year.

Plumber's Licenses

You need a license from the State Plumbing Board to do plumbing work in Louisiana. To get an application, contact:

State Plumbing Board of Louisiana
2714 Canal Street, Suite 512
New Orleans, LA 70119
(504) 826-2382
Fax: (504) 826-2175

The Board issues master and journeyman licenses. You must pass an exam to get a license. To qualify for the master exam you must have a journeyman license in Louisiana. To qualify for the journeyman exam you must verify at least five years or 8,000 hours of work experience installing, repairing, or maintaining plumbing systems. The Board doesn't have any reciprocity agreements with other states but if you have a valid plumbing license in another state you can use it to qualify to take the journeyman plumber exam in Louisiana.

Plumber's license fees: It will cost you $100 to take the master exam and the journeyman exam is $75. The master license costs $180 and the journeyman license is $30. All licenses are good for one year, expiring on December 31 each year.

Water Well Driller's License

Except for oil and gas wells you must get a license from the Louisiana Department of Transportation and Development to do well drilling work in Louisiana. To get an application, contact:

Water Resource Section
Department of Transportation and Development
P.O. Box 94245
Baton Rouge, LA 70804-9245
(504) 379-1434
Fax: (504) 379-1857

You need to pass an exam and have two years of approved well drilling experience to qualify for a license. You must send two letters of reference with your application.

You will have to appear in person before the Advisory Committee for Regulation and Control of Water Well Contractors and be interviewed by the Committee. They meet on the first Wednesday of February, May, August, and November each year. After the Committee approves your application, you can take their exam.

Water well driller's license fees: It will cost you $100 to get a license. The exam costs $10. All licenses expire each June 30.

If you have a valid driller's license from another state you may be able to get a Louisiana driller's license. The requirements for your license must be approximately equal to those for a Louisiana driller license and the state you're licensed in must allow a Louisiana driller license holder to get a license in the state the same way. Currently the Department doesn't have any reciprocity agreements with other states.

Department of Transportation (DOT)

You don't need to be prequalified to bid on contracts from the Department of Transportation. However you must be licensed by the State Licensing Board for Contractors. For information on this see our discussion at the beginning of the Louisiana section.

Out-of-State Corporations

Out-of-state corporations must get a Certificate of Authority to do business in Louisiana from the Louisiana Secretary of State. To apply for this certificate, contact:

Contact

Louisiana Secretary of State
Corporation Division
P.O. Box 94125
Baton Rouge, LA 70804
(504) 925-4704
Fax: (504) 925-4726

Maine

General building contractors don't need a license in Maine. You'll need to be licensed to do asbestos abatement work, or electrical or plumbing contracting.

Asbestos Abatement Contractor's Licenses

If you work with asbestos in Maine you must be certified and/or licensed by the Maine Department of Environmental Protection. The Department issues these asbestos abatement contractor's licenses:

Full

In-house (limited to owner properties)

Conditional (limited to roofing or siding)

The Department also issues certificates to these asbestos trades:

Worker	Air monitor
Supervisor	Air analyst
Inspector	Bulk analyst
Design consultant	Management planner

To get an application for a certificate or license contact:

Contact

State of Maine
Department of Environmental Protection
17 State House Station
Augusta, ME 04333-0017
(207) 287-2651
web site: www.state.me.us/dep

To qualify for a contractor's license you must:

1. Have access to at least one state-approved asbestos disposal site

2. Have a worker protection and medical monitoring plan approved by EPA, OSHA, and/or Code of Federal Regulations

3. Be familiar and comply with all applicable state and federal standard for asbestos abatement projects

4. Provide company standard operating training for all employees before starting a job

5. Be capable of complying with all state, EPA, OSHA, and Code of Federal Regulation rules

6. Provide a list of all personnel (to include at least one certified Supervisor) and asbestos equipment

7. Provide a list of states you are licensed in for asbestos work

8. Provide a list of any citations received in last two years for asbestos work

9. Provide proof that all employees have been certified by the Department

You must also make sure all employees who come in contact with asbestos or who will be responsible for an asbestos abatement project have fulfilled items 3, 4, and 5 above.

Contractor's license fees: The contractor's license will cost you $250 nonrefundable and it's good for a year.

If you can prove you have a valid license in a state with qualifications at least equal to Maine's, you can apply for a Department license. You still need to pay the fee though.

To qualify for a trade certificate you need to submit a signed statement saying that you know and will comply with your employer's standard operating procedures. You also need to pass a Department-approved training course which includes an exam. A certificate is good for one year.

Here's a table showing the length of the Department-approved course and the nonrefundable fee you must pay to apply for an asbestos trade certificate:

Trade	Course hours	Fee
Worker	32	$25
Supervisor	40	$50
Air monitor	16	$50
Inspector	24	$50
Management planner	16	$50
Design consultant	24	$50
Air analyst	36	$50
Bulk analyst	36	$50

If you can prove you have a valid certificate in a state with qualifications at least equal to Maine's, you can apply for a Department certificate. You still need to pay the fee though.

Electrician's Licenses

Electrical work in Maine requires a license. Maine issues these electrician licenses: journeyman, master and limited. The limited electrical trades are:

Water pumps
Outdoor signs
Gasoline dispensing
Traffic signals

House wiring
Refrigeration
Low energy

To apply for a license contact:

Contact

Electricians' Examining Board
35 State House Station
Augusta, ME 04333-0035
(207) 624-8610
Fax: (207) 624-8637

For the master's license you'll need 12,000 hours of work experience or 4,000 hours working as a journeyman. For the journeyman license you'll need 8,000 hours of work experience, a combination of work experience and graduation from an accredited technical school, or work experience and an apprenticeship program. For either a master's or journeyman's license you need to complete 576 hours of study approved by the Board and a course that's at least 45 hours long on the current *National Electrical Code*.

Here are the requirements for the limited licenses:

Limited trade license	Board-approved course hours	Work experience
Water pumps	135	2,000
Outdoor signs	135	2,000
Gasoline dispensing	135	2,000
Traffic signals	135	2,000
House wiring	225	4,000
Refrigeration	270	6,000
Low energy	270	4,000

The Board will review your application and if you're eligible they'll send you information on the exam they require. National Assessment Institute gives the exam. For information on the exam or scheduling, you can contact NAI at:

Contact

National Assessment Institute
2 Mount Royal Avenue, Suite 250
Marlborough, MA 01752
(508) 624-0826

Here are the subjects on the journeyman and master exams:

	Percent of exam	
Subject	**Journeyman**	**Master**
General knowledge of the electrical trade and calculations	17	25
Raceways and enclosures	14	11
Services, feeders, and branch circuits	12	13
Grounding and bonding	11	11
Conductors	11	9
Utilization and general use equipment	11	11
Special occupancies/equipment	10	6
Motor and controls	9	11
Alarm, communication and low voltage circuits	5	3

Fee

Electrician's license fees: It will cost you $20 to file an application for a license and $50 for an exam. The master license costs $150, $80 for the journeyman, and $100 for the limited trade license. Licenses are good for two years.

If you have a valid license in a state with qualifications at least equal to Maine's, six years of work experience, and the current 45-hour *National Electrical Code* course, you may be able to get a Maine electrician's license without taking an exam. You still need to pay the fees.

Recommended Reading for the Electrical Exams

📖 *NFPA 70 - National Electrical Code,* 1996 edition, National Fire Protection Association, 1 Batterymarch Park, P.O. Box 9101, Quincy, MA 02269

📖 *American Electricians Handbook,* 1996, 13th edition, Croft/Summers, McGraw-Hill Inc., Box 543, Blacklick, OH 43004-0543

📖 *Sign Electricians' Workbook,* 1993, James G. Stallcup, American Technical Publishers, Inc.

📖 *Design and Application of Security/Fire Alarm Systems,* John Traister, McGraw-Hill Company

📖 *NEMA Training Manual on Fire Alarm Systems,* National Electrical Manufacturers Association

📖 *NFPA 72-1993, National Fire Alarm Code,* National Fire Protection Association, 1 Batterymarch Park, P.O. Box 9101, Quincy, MA 02269

📖 *Proprietary Burglar Alarm Systems,* U.L. #1076, Underwriters Laboratories, Inc.

📖 *Understanding and Servicing Alarm Systems,* Trimmer and Butterworth, Butterworth Publishers

---

Recommended Reading for the Electrical Exams (continued)

📖 *Fire Alarm Signaling Systems Handbook,* Bukowski and O'Laughlin, National Fire Protection Association

📖 *Neon Installation Manual,* International Association of Electrical Inspectors

You can get these books from:

Builders' Book Depot
1033 Jefferson, Suite 500
Phoenix, AZ 85034
(800) 284-3434

Plumber's Licenses

Plumbing work in Maine requires a license. Maine issues journeyman, master, and trainee licenses. To apply for a license, contact:

Contact

Plumbers' Examining Board
35 State House Station
Augusta, ME 04333-0035
(207) 624-8603
Fax: (207) 624-8637

For a master's license you need at least one year with 2,000 hours of work experience as a journeyman or four years with 8,000 hours of work as a trainee plumber. For the journeyman license you need at least two years with 4,000 hours of work experience as a trainee plumber.

The Board will review your application and if you're eligible they'll send you information on the exam they require (no exam for a trainee license). The exam is closed book with 100 multiple choice questions. It's three hours long. For information on the exam or scheduling, you can contact NAI at:

Contact

National Assessment Institute
2 Mount Royal Avenue, Suite 250
Marlborough, MA 01752
(508) 624-0826

Here's the content of the exam:

Subject	Percent of exam
General code knowledge	22
Installation practices, methods, materials	16
Drains and vents, sewers	16
Water supply, backflow prevention	16

Subject	Percent of exam
Special wastes, roof drains	10
Fixtures, trim	5
Excavation	5
Inspection, testing	5
Alternate systems of plumbing	5

Fee

Plumber's license fees: It will cost you $25 to file an application for a license and $60 for an exam. The master license costs $150, $75 for a journeyman, and $40 for a trainee. All master licenses are good for two years, expiring on October 31. Other plumbing licenses are also good for two years, expiring on April 30.

If you have a valid license in a state with qualifications at least equal to Maine's, you may be able to get a Maine license by filing out an application, supplying a certified copy of your license to Maine, and paying the necessary fees.

Recommended Reading for the Plumbing Exams

📖 *State of Maine Plumbing Code (Internal and External),* Department of Human Services 10 State House Station, Augusta, ME 04333-0010

📖 *Plumbing,* L. V. Ripka, American Technical Publishers, 1155 West 175th Street, Homewood, IL 60403

Underground Tank Installer's Certificate

To safeguard public health, safety and the environment, Maine certifies tank installers and removers. However you don't have to be certified to remove tanks that store fuels the National Fire Protection Association says aren't explosive.

Only an individual can be certified in Maine, not a business or corporation. To apply for certification, contact:

Contact

Department of Environmental Protection
State Board of Underground Storage Tank Installers
State House Station #17
Augusta, ME 04333-0017
(207) 287-2651
Fax: (207) 287-7826

The Board issues four types of certificates:

- Underground gasoline tank remover
- Underground hazardous substance tank installer
- Class 2 storage tank installer (may install or remove any type of underground oil storage tank except field-constructed storage or impressed-current cathode-protected tanks)
- Class 3 storage tank installer (may only install or remove underground storage tanks for #2 heating oil)

Fee

Underground tank installer's certificate fees: Let's look first at the first two types of certificates. Both of these require you to pass a multiple choice exam with 100 questions that the Board has approved. The fee for the Hazardous Substance Tank Installer exam is $50 nonrefundable, and $25 nonrefundable for the Tank Remover exam. You need a score of at least 80% to pass. The Board also requires you to demonstrate your skill by installing (or removing) a tank in the field. You have six months after you pass the written exam to do this. The Board will check your work and decide if you pass or not. This exam will cost you $350 nonrefundable for a one-day project. If you need more time than that, it'll cost you $350 per day more. You only have one chance to pass the on-site exam. If you fail this exam, you have to pass the written exam again before you can try the on-site exam another time. When you pass the on-site exam and pay a certification fee of $150 you'll get your certificate.

If you have a certificate in a state with qualifications at least equal to Maine's, you may be able to get a Maine certificate by filing out an application, supplying a certified copy of your certificate to the Board, and paying a $250 nonrefundable reciprocity fee.

Getting a Class 2 or 3 Installer certificate has more steps. First you file your application and pay a $200 nonrefundable fee. Then you have to score at least 80% on a Board-approved 100-question exam. This qualifies you to be an apprentice installer. As an apprentice for a Class 2 certificate you must complete six documented installations of tanks used for motor fuel or marketing and distributing oil. For a Class 3 certificate you need six completed heating oil installations (also documented). If you've assisted on at least that many installations outside Maine, you may be able to get the Board to give you a variance here. You'll have to pay a $200 nonrefundable fee for processing of this variance.

After you complete these installations you have to take a Board-approved written final exam. This exam costs another $250 nonrefundable fee. When you pass this exam by scoring at least 80% and pay a certification fee of $150, you'll get your certificate.

If you have a certificate in a state with qualifications at least equal to Maine's, you may be able to get a Maine certificate by filing out an application, supplying a certified copy of your certificate to the Board, and paying a $250 nonrefundable reciprocity fee.

All of these certificates are good for two years.

Department of Transportation (DOT)

The Maine Department of Transportation doesn't prequalify contractors to bid on Department projects.

Out-of-State Corporations

Out-of-state contractors working in Maine must be licensed by the Division of Corporations. For information contact:

Contact

Bureau of Corporations
Secretary of State
101 State House Station
Augusta, ME 04333-0101
(207) 287-4190
Fax: (207) 287-5874

Maryland

General construction contractors don't need a license to work in Maryland. You will need a license to do electrical, plumbing or HVACR contracting, or work on home improvement projects.

Electrician's Licenses

To do electrical work you must get a master electrician license from the Maryland Board of Master Electricians. Unless you qualify for a waiver, you must pass an exam given by PSI Examination Services before you can get an application from the Board. You'll find information on exam waivers later in this section. To get an application for the exam, contact:

Contact

PSI Examination Services
100 West Broadway, Suite 1100
Glendale, CA 91210-1202
(800) 733-9267
Fax: (818) 247-3853

To qualify for the exam you must have seven years of work experience under the supervision of a master electrician or similarly-qualified person. If you have Board-approved professional training, you can apply up to three years of it toward the experience requirement.

The exam is open book and lasts four hours. It has 50 one-point questions, 10 two-point questions, and 6 five-point questions. Here's a table that shows the subjects on the exam and the approximate number of questions on each subject:

Subject	Approximate number of questions
General theory	10 2-point
Wiring and protection	14 1-point
Wiring methods, materials	14 1-point
Equipment for general use	8 1-point

Subject	Approximate number of questions
Special occupancies	8 1-point
Special equipment	4 1-point
Special conditions	2 1-point
Applications and calculations	6 5-point

The reference for the exam is the *National Electric Code*.

After you pass the exam you can apply to the Board for your license. Here's their address:

Contact

State of Maryland
Board of Master Electricians
501 St. Paul Place, Room 901
Baltimore, MD 21202-2272
(410) 333-6322
Fax: (410) 333-6314

Fee

Electrician's license fees: It will cost you $40 nonrefundable to apply for the exam and $75 for a master license. Licenses expire on June 30 each odd-numbered year.

If you have a local or out-of-state master electrician's license, you may be able to get a master's license from the Board without taking the exam. Your license must be active and in good standing. The requirements you filled to get it must be essentially equal to those required to get a Maryland license. You'll still have to pay the license fee but you can apply directly to the Board for your license.

The Board doesn't have any reciprocity agreements with other states at this time.

Plumber's Licenses

To do plumbing work you must get a master or journey license from the Maryland State Board of Plumbing. Unless you qualify for a waiver, you've got to pass an exam given by PSI Examination Services before you can get an application from the Board. To get an application for the exam contact:

Contact

PSI Examination Services
100 West Broadway, Suite 1100
Glendale, CA 91210-1202
(800) 733-9267
Fax: (818) 247-3853

To qualify for the master or journey exam you must have completed a Board-approved 32-hour course in Backflow Prevention Devices.

The Board has different requirements for the master license, depending on whether your work experience was in Maryland or not. For work experience in Maryland you need two years Board-approved work experience under a licensed Maryland master plumber, including at least 3,750 hours in the plumbing trade as a licensed Maryland journey plumber.

For work experience outside Maryland, you need:

- two years of Board-approved work experience including at least 3,750 hours in the plumbing trade as a licensed journey plumber
- four years of Board-approved work experience including at least 7,500 hours under the supervision of a licensed master plumber before getting your journey plumber license
- proof that you got your journey plumber license by passing a written exam
- 11,250 hours of Board-approved work experience under the supervision of a licensed master plumber who got his license by passing a written exam

The requirements for taking the journey plumbing licensing exam are also different depending on whether your work experience was in Maryland or not. For work experience in Maryland you need four years of Board-approved work experience under a licensed Maryland master plumbing, including at least 7,500 hours under the supervision of a licensed Maryland master plumber.

You can apply up to 1,500 hours of Board-approved education to the 7,500-hour requirement. You must also list any hours you've worked as an apprentice plumber. Any hours you work as an unlicensed apprentice plumber will be subtracted from your hours worked as an apprentice plumber. An unlicensed apprentice plumber must register with the Board before applying for the journey plumber exam.

For work experience outside Maryland, you need:

- four years of Board-approved work experience including at least 7,500 hours under the supervision of a licensed master plumber who got his license by passing a written exam

You can apply up to 1,500 hours of Board-approved education to the 7,500-hour requirement.

The exam is open book on the *Maryland State Plumbing Code* and the *National Fuel Gas Code*. Here are the subjects on each exam and their approximate number of questions:

	Approximate number of questions	
Subject	**Journey exam**	**Master exam**
Plumbing fundamentals	10	7
Materials use, specifications	8	8
Traps, interceptors, backwater valves	5	5
Plumbing fixtures	5	5

Subject	Approximate number of questions	
	Journey exam	Master exam
Water supply systems	10	10
Backflow prevention	10	10
Drain, vent, waste systems	15	15
Special topics	0	3
Gas appliances, piping	10	10
Plumbing mathematics	10	10

After you pass the exam you can apply to the Board for your license. Their address is:

Maryland State Board of Plumbing
501 St. Paul Place, Room 900
Baltimore, MD 21202-2272
(410) 333-6328
Fax: (410) 333-6314

Plumber's license fees: It will cost you $50 nonrefundable to apply for the exam. A master license costs $100 and a journeyman license is $50. Licenses expire on May 1 each odd-numbered year.

At this time, the only reciprocity agreement the Board has is with the Washington Suburban Sanitary Commission.

HVACR Licenses

To do HVACR work you must get a license from the Maryland State Board of HVACR Contractors. The Board issues master, master restricted, limited contractor, and journeyman licenses. To get an application, contact:

State Board of HVACR Contractors
500 N. Calvert Street
Baltimore, MD 21202-3651
(410) 333-6590
Fax: (410) 333-6314

HVACR license fees: You must pass an exam to get a license. The master or limited contractor exam costs $132. The fee for the master restricted exam is $44 for one area (forced air heating, hydronic heating, ventilation, air conditioning or refrigeration), and $88 for two or more areas. The journeyman exam costs $44. The master license costs $150, limited contractor costs $125, master restricted is $50, and the journeyman license costs $35.

Home Improvement Licenses

To work on home improvement projects you must get a contractor or subcontractor license from the Maryland Home Improvement Commission. But before you can get the application, you must pass an exam given by National Assessment Institute (NAI). To get an application for the exam, contact NAI at:

Contact

National Assessment Institute
3823 Gaskins Road
Richmond, VA 23233
(800) 356-3381

To qualify for an exam you must have two years of work experience. At least one year must be working in a trade. You can apply one year of Commission-approved education or Commission-approved work as a supervisor, manager, or owner to the work experience requirement.

The exams are open book. The contractor exam has 75 multiple choice questions and lasts three hours. The subcontractor exam has 50 questions and lasts two hours. Here are the subjects on each exam and the approximate percentage of questions of each subject:

	Approximate % of exam	
Subject	**Contractor**	**Subcontractor**
Home improvement law	45	50
Door-to-door sales act	13	5
Labor laws	10	15
Safety regulations	7	8
Payroll taxes	7	8
Statutory liens of real property	3	4
Business, finances, estimating	15	20

The reference for the exams is the *Reference Manual for Home Improvement*, which is published every year. You can get a copy from NAI for $26.90.

After you pass the exam you'll get an application to fill out for your license. You'll need to have liability insurance for at least $50,000, copies of your last three personal bank statements, and an unaltered, original credit report that's less than 60 days old. If your company is a corporation, you need corporation papers and a letter of good standing from the State Department of Assessments and Taxation.

Fee

Home improvement license fees: It will cost you $55 nonrefundable to apply for an exam and $325 (includes Home Improvement Guaranty Fund fee) for a contractor license. A subcontractor license costs $125. All licenses expire June 30 each odd-numbered year.

Department of Transportation (DOT)

You don't need to be prequalified to bid on Maryland Department of Transportation projects. However you do need an appropriate contractor's license. If you're the probable low bidder on a project, the Department will ask you about your organization, personnel, financial condition, equipment, and experience at that time. For information, contact:

Contact

Department of Transportation
707 North Calvert Street, Room 507
Baltimore, MD 21203
(410) 545-0433

Out-of-State Corporations

Out-of-state corporations must file a Foreign Qualification form. To get this form, contact:

Contact

State Department of Assessments and Taxation
301 W. Preston Street, Room 809
Baltimore, MD 21201
(410) 767-1006
Fax: (410) 333-7097

Massachusetts

If you supervise anyone (including yourself) in construction work or demolition in Massachusetts, you need a license. There are two possibilities for this license — unrestricted and restricted. The Restricted Construction Supervisor's License limits you to supervising work on one- and two-family dwellings and any of their accessory buildings. The Unrestricted License limits you to work on building structures less than 35,000 cubic feet. To apply for either license, contact:

Contact

State Board of Building Regulations and Standards
McCormack State Office Building
One Ashburton Place, Room 1301
Boston, MA 02108
(617) 727-3200
Fax: (617) 227-1754

You must pass an exam to be licensed, and you must have at least three years of full time work experience to take the exam. The exam is three hours long, open book with 50 multiple choice questions. It covers:

Subject	Number of questions
Code administration	7 - 9
Quality, strength, safety of materials and designs	3 - 5
Site work and foundations	3 - 5
Concrete	3 - 5
Masonry	3 - 5
Metals	2 - 4
Wood	7 - 9
Thermal and moisture protection	4 - 6
Doors, windows, and passages	2 - 4
Finishes	0 - 2
Mechanical systems	2 - 4
Fire protection	2 - 4

The testing company has developed a Home Study Workbook for the exam. You can get a copy from them for $37. For information on the test, contact:

Contact

Massachusetts Construction Supervisors Licensing Examinations
Educational Testing Service
506 Carnegie Center
P.O. Box 6530
Princeton, NJ 08541-6530
(609) 921-9000

Fee

Contractor's license fees: The exam costs $75. The license costs $150.

Home Improvement Contractor's Licenses

If you work on existing residential structures that are one- to four-unit owner-occupied buildings, you also need to register with the state. To do this, contact the State Board of Building Regulations and Standards (address on page 141) and ask for an application to register as a Home Improvement Contractor or Subcontractor. There's no exam connected with this registration.

Home Improvement Contractor's registration will cost you $100. But you may not have to it this if you have a Construction Supervisor license in Massachusetts.

There's also a Guaranty Fund that you may have to contribute to the first time you register. The amount you need to contribute will depend on how many employees your company has:

Number of employees	Required fund contribution
0 - 3	$100
4 - 10	$200
1 - 30	$300
More than 30	$500

Asbestos or Lead Abatement Licenses

To do asbestos or lead abatement work in Massachusetts, you need a license. To get an application, contact:

Contact

Department of Labor & Workforce Development
Division of Occupational Safety
100 Cambridge St., Room 1106
Boston, MA 02202
(617) 727-7047
(800) 425-0004 (in Massachusetts only)

Here are the licenses the Department issues and the nonrefundable fee for each one:

License	Fee
Deleader supervisor	$125
Deleader worker	—
Asbestos supervisor	$100
Asbestos worker	$25

Before you can get any of these licenses you must have completed the training the Department requires.

Electrician's Licenses

To do electrical work in Massachusetts you need a license. All electrical licenses are good for three years. To get an application, contact:

Division of Registration
Board of State Examiners of Electricians
100 Cambridge St., Room 1511
Boston, MA 02202
(617) 727-9931

The Board issues journeyman, master, system contractor, and system technician licenses. You must pass an exam for any of these licenses. To take the journeyman exam you need three years of electrical work experience (four years if you started in the profession after September 1, 1981), and 300 hours of Board-approved education. You need to hold a journeyman license from the Board at least one year before you can take the master exam. Both exams are open book with multiple choice questions. The journeyman exam last three hours and has 60 questions. The master exam lasts four hours and has 80 questions. Here's a summary of the test requirements:

	Percent of exam	
Subject	Journeyman	Master
General knowledge of the electrical trade and calculations	18	25
Raceways and enclosures	14	11
Services, feeders, and branch circuits	14	13
Grounding and bonding	11	11
Conductors	12	9
Utilization and general use equipment	11	11
Special occupancies/equipment	6	6
Motor and controls	11	11
Low voltage circuits including alarms and communications	3	3

You also need to pass a practical exam to get a journeyman license. This exam will ask you to identify pictures and drawings of electrical materials and answer questions on the Massachusetts electrical laws and code amendments. The exam has 23 questions. Here's a breakdown of its contents:

Subject	Approximate number of questions
Electrical diagrams	5
Identifying electrical materials	10
State code amendments	5
State licensing laws	3

You can take this exam the same day you take the journeyman exam.

If you have a journeyman or master electrician license in New Hampshire, Vermont, Washington, Maine, or Oregon you can apply to get a license from the Board by reciprocity.

To take the system technician exam you need three years of work experience under the supervision of a licensed systems technician and 200 hours of Board-approved education. You need to hold a system technician license from the Board at least one year and work at least one year at the trade before you can take the system contractor exam. You also need 75 hours of Board-approved course work in advanced systems technology and business management.

The exams for these licenses are open book, multiple-choice format. The system contractor exam has 60 questions and lasts four hours. The technician exam has 50 questions and last three hours. The exams are on:

National Electrical and Massachusetts Codes (Articles 90, 100, 110, 250, 300, 310-Table, 310-316, 318, 330-337, 334-352, 354, 356, 358, 362, 370, 480, 500, 700-702, 720, 725, 760, 770, 800, Chapter 9, Part A8)

General technical knowledge

  Equipment
  Signals transmission
  Systems design considerations

All the electrical licensing exams are given by National Assessment Institute (NAI). You can contact them for information on the exams at:

Contact

National Assessment Institute
2 Mount Royal Avenue, Suite 250
Marlborough, MA 01752
(508) 624-0826

Here are the licenses available, and fees for each:

License	Nonrefundable application fee	Exam fee	License fee
Journeyman	$12	$80	$65
Journeyman (written part only)	$12	$60	—
Journeyman (practical part only)	$12	$40	—
Master	$25	$60	$85
Systems contractor	$25	$60	$115
Systems technician	$12	$80	$85
Systems technician (written part only)	$12	$60	—
Systems technician (practical part only)	$12	$40	—

Recommended Reading for the Electrical Exams

Master and Journeyman Exams:

- *NFPA 70 - National Electrical Code*, 1996 edition, National Fire Protection Association, 1 Batterymarch Park, P.O. Box 9101, Quincy, MA 02269

- *Alternating Current Fundamentals*, Duff and Herman, Delmar Publishers, P.O. Box 6904, Florence, KY 41022

- *Direct Current Fundamentals*, Loper and Tedsen, Delmar Publishers, P.O. Box 6904, Florence, KY 41022

- *Electric Motor Control*, Walter N. Alerich, Delmar Publishers, P.O. Box 6904, Florence, KY 41022

- *National Electrical Code Blueprint Reading*, K. L. Gebert, American Technical Publishers, 1155 West 175th Street, Homewood, IL 60403

- *American Electricians' Handbook*, 12th edition, Croft, Watt, and Summers, McGraw-Hill Company

- *Ferm's Fast Finder Index*, Olaf G. Ferm, Ferms Finder Index Company

System Technician Exam:

- *Fire Alarm Signaling Systems Handbook*, Bukowski and O'Laughlin, National Fire Protection Association, 1 Batterymarch Park, P.O. Box 9101, Quincy, MA 02269

- *Understanding and Servicing Alarm Systems*, Trimmer and Butterworth, Butterworth-Heinemann Publishers

You can get the reference books from:

Builders' Book Depot
1033 Jefferson, Suite 500
Phoenix, AZ 85034
(800) 284-3434

Department of Transportation (DOT)

You must be prequalified to bid on Massachusetts Department of Highways work. Prequalification is good for up to 24 months. To get an application, contact:

Massachusetts Highway Department
Prequalification and Contract Office
10 Park Plaza, Room 7373
Boston, MA 02116-3973
(617) 973-7620

The application will ask you:

- what construction equipment you own
- what your work experience is
- what work you've completed
- what surety company will furnish your bonds
- what classes of work you want to be prequalified in

Here are classes of work the Highway Department uses:

Asbestos removal	Electrical lighting
Bridge construction	Marine construction
Catch basin cleaning	Mowing and spraying
Chemical storage sheds	Structural painting
Crack sealing	Pavement marking
Demolition	Processing, recycling, transporting excavated soils
Drawbridge maintenance	
Drilling and boring	Reclamation
Impact attenuators	Sewer and water
Intelligent transportation systems	Structural signs
Guardrail and fencing	Surfacing
Hazardous waste remediation	Traffic signals
Highway construction	Tunnels
Highway maintenance	Underground tank removal and replacement
Highway sweeping	
Landscaping and roadside development	Utilities

Out-of-State Corporations

You must register with the Massachusetts Secretary of State and get a Certificate of Good Standing to do business in the state. To get an application, contact:

Office of the State Secretary
One Ashburton Place, Room 1717
Boston, MA 02108
(617) 727-9640

Michigan

To work on residential or a combination of residential and commercial buildings in Michigan, you need to be licensed by the Builder's Unit of the Department of Consumer and Industry Services. The Department issues Residential Builder and Maintenance and Alteration Contractor's licenses. But before you can apply for a license, you must first pass an exam. The exam is given by PSI Examination Services. So you need to get an application for an exam from them. You can contact them at:

Contact

PSI Examination Services
100 West Broadway, Suite 1100
Glendale, CA 91210
(800) 733-9267
Fax: (818) 247-3853

Residential Builder's License

Let's look first at the Residential Builder license. This license will let you work in these trades:

Basement waterproofing	Carpentry
Concrete	Excavation
Gutters, downspouts	House wrecking
Insulation	Masonry
Painting, decorating	Roofing
Siding	Storms, screens
Swimming pools	Tile, marble

This license doesn't let you do plumbing, electrical, or mechanical work. See below for information on those trades.

You must pass a two-part exam to apply for this license — a business and law part and a trade part. Each part has 50 multiple choice questions and lasts one and one-half hours. Here are the subjects on the business and law part and the number of questions on each subject:

Subject	Number of questions
Regulator and statutory requirements	20
Financial management	8
Contract management	6
Project management	7
Safety, personnel, payroll	4
Insurance, bonding, liens	4
Business organizations	1

Here's a summary of the subjects on the trade part and the number of questions on each subject:

Subject	Number of questions
Excavation and site work	7
Rough carpentry	16
Finish carpentry	13
Concrete and rebar	6
Roofing	3
Masonry	2
Diverse specialties	3

Recommended Reading for Residential Builder's License

Business and Law:

- *BOCA National Building Code,* Building Officials and Code Administrators International Inc., 1993, Michigan Department of Consumer and Industry Services, Office of Management Services, Bureau of Construction Codes, P.O. Box 30255, Lansing, MI 48909

- *Michigan Construction Code Commission General Rules Part 4 - Building Code Rules,* Michigan Department of Consumer and Industry Services, Office of Management Services, Bureau of Construction Codes, P.O. Box 30255, Lansing, MI 48909

- *Michigan Occupational Safety and Health Act 154 of 1974,* Michigan Department of Consumer and Industry Services, Safety Standards Division, P.O. Box 30015, Lansing, MI 48909

Recommended Reading for Residential Builder's License (continued)

Business and Law (continued):

📖 *Michigan Workers' Compensation Law, Michigan Department of Consumer and Industry Services,* Bureau of Workers' Disability Compensation, P.O. Box 30016, Lansing, MI 48909

📖 *Handy Reference Guide to the Fair Labor Standards Act,* 1994, U.S. Department of Labor, 14th Street and Construction Avenue, Washington, DC 20210

📖 *Michigan Single Business Tax Booklet for Individuals, Partnerships, and Fiduciaries,* State of Michigan, Department of Treasury, Treasury Building, P.O. Box 15128, Lansing, MI 48901

📖 *Circular E Employer's Tax Guide,* 1993, Internal Revenue Service

📖 *Fair Housing in Michigan: Michigan Civil Rights Act: Article 5, Michigan Civil Rights Commission,* Victor Office Center, Suite 700, 201 N. Washington Square, Lansing, MI 48913

📖 *Residential Builders Laws and Rules Relating to Residential Builders and Maintenance and Alteration Contractors,* State of Michigan Residential Builders and Maintenance and Alteration Contractor Board

📖 *Construction Contracting,* 6th edition, 1994, R. Clough, John Wiley & Sons, Inc., 605 Third Avenue, New York, NY 10158

📖 *Builder's Guide to Accounting,* 1987, Thomsett, Craftsman Book Company, 6058 Corte del Cedro, P.O. Box 6500, Carlsbad, CA 92018

Trade Exam:

📖 *BOCA National Building Code,* Building Officials and Code Administrators International Inc., 1993, Michigan Department of Consumer and Industry Services, Office of Management Services, Bureau of Construction Codes, P.O. Box 30255, Lansing, MI 48909

📖 *Michigan State Construction Code Act 230 of 1972,* Michigan Department of Consumer and Industry Services, Office of Management Services, Bureau of Construction Codes, P.O. Box 30255, Lansing, MI 48909

📖 *Construction Safety Standards,* Michigan Department of Consumer and Industry Services, Safety Standards Division, P.O. Box 30015, Lansing, MI 48909

📖 *Excavation & Grading Handbook,* 1991, Nicholas E. Capachi, Craftsman Book Company, 6058 Corte del Cedro, P.O. Box 6500, Carlsbad, CA 92018

📖 *New Complete Do-It-Yourself Manual,* 1991, Reader's Digest, The Reader's Digest Association, Inc., Pleasantville, NY 10570

Recommended Reading for Residential Builder's License (continued)

Trade Exam (continued):

📖 *Modern Carpentry,* 1992, Willis H. Wagner, The Goodheart-Willcox Company, Inc., 123 West Taft Drive, South Holland, IL 60473

📖 *Design and Control of Concrete Mixtures,* 1988/1990, 13th edition, Portland Cement Association, 5420 Old Orchard Road, Skokie, IL 60077-1083

📖 *Carpentry and Building Construction,* 1993, Feirer, Hutchings & Feirer, McGraw-Hill Inc., Box 543, Blacklick, OH 43004-0543

📖 *Concrete Form Construction,* 1977, Moore, Delmar Publishers, P.O. Box 6904, Florence, KY 41022

📖 *Concrete Masonry Handbook for Architects, Engineers, Builders,* 1985, Randal & Panarese, Portland Cement Association, 5420 Old Orchard Road, Skokie, IL 60077-1083

📖 *NRCA Roofing and Waterproofing Manual,* 1996, 4th edition, National Roofing Contractor's Association, 10255 W. Higgins, Rd., Suite #600, Rosemont, IL 60018-5607

📖 *Roofers Handbook,* 1992, 8th printing, William Johnson, Craftsman Book Company, 6058 Corte del Cedro, P.O. Box 6500, Carlsbad, CA 92018

📖 *Walker's Insulation Techniques and Estimating Handbook,* 1983, H. Hardenbrook, F. Walker Co., 5030 North Harlem Ave., Chicago, IL 60656

📖 *Painting and Decorating Encyclopedia,* 1982, W. Brushwell, The Goodheart-Willcox Company, Inc., 123 West Taft Drive, South Holland, IL 60473

📖 *Handbook for Ceramic Tile Installation,* 1995, 33rd edition, Tile Council of America, P.O. Box 326, Princeton, NJ 08542-0326

📖 *Building Trades Blueprint Reading - Part 2,* 1987, Sundberg and Proctor, American Technical Publishers, 1155 West 175th Street, Homewood, IL 60403

Maintenance and Alteration Contractor's License

Now let's look at the Maintenance and Alteration Contractor license. Like the Residential Builder license, you must pass the business and law part of the exam to apply for this license. However the Maintenance and Alteration Contractor license lets you work only in the trades you pass a trade exam in. You can be licensed in as many trades as you like but you must pass an exam for each one. Each trade exam has 25 to 50 multiple choice questions and lasts one and one-half hours. Here are the subjects on the trade exams and the number of questions on each:

Trade	Number of questions
Basement waterproofing	40
Carpentry	50
Concrete	50
Excavation	40
Gutters, downspouts	35
House wrecking	25
Insulation	50
Masonry	50
Painting, decorating	50
Roofing	50
Siding	30
Storms, screens	35
Swimming pools	50
Tile, marble	50

This license also doesn't let you do plumbing, electrical, or mechanical work. See below for information on those trades.

Maintenance and alteration contractor's license fees: It will cost you $53 nonrefundable to take the business and law part of the exam and one of the following, depending on which license you're applying for:

- one trade for the Residential Builder license
- one or two trades for the Maintenance and Alteration Contractor license

It will cost you $30 to add more trades to the Maintenance and Alteration Contractor license.

After you pass all the required exams, it will cost you $95 for either license. A license is good for two years. The license will come from the Builder's Unit of the Department of Consumer and Industry Services. You can contact them at:

Builder's Unit
Office of Commercial Services
Department of Consumer and Industry Services
P.O. Box 30245
Lansing, MI 48909

Electrician's Licenses

To do electrical work in Michigan you need a master or journeyman license. To get an application for a license, contact:

Contact

Michigan Department of Labor
Bureau of Construction Codes
Electrical Administrative Board
7150 Harris Drive
P.O. Box 30255
Lansing, MI 48909
(517) 241-9320
Fax: (517) 241-9308

You'll have to pass an exam to get a license. To qualify for the master exam you need:

- six years of electrical work experience covering 12,000 hours
- two years of work experience as a licensed journeyman electrician

If you're licensed in another state which has licensing requirements about the same as Michigan, you can qualify to take the exam. Also if you don't fulfill these requirements you can still present evidence of your experience to the Board and possibly get permission to take the exam.

The master exam covers knowledge required to plan and supervise electrical installation, read plans and drawings, and apply knowledge of relevant safety procedures and requirements. Here are the subjects on the exam:

Grounding and bonding

Branch circuits, wire connections and devices

Conductors

General knowledge of electrical trade

Motors and control of motors and equipment

Services and feeders

General use equipment

Overcurrent protection

Raceways

Special occupancies and equipment

Boxes, cabinets, panelboards, non-raceway enclosures

Low voltage circuits and equipment

Lighting and lamps

State laws, rules, code amendments

Fee

Electrician's license fees: It will cost you $25 to apply for the master exam. The license costs $25 and it's good for one year, expiring on December 31 each year. If you can prove you have a valid license in a state with qualifications at least equal to Michigan's, you may be eligible for a Michigan license without taking the exam. You'll still need to pay the fee.

To qualify for the journeyman exam you must have 8,000 hours of electrical work experience over at least four years working under a master electrician. If you're licensed in another state which has licensing requirements about the same as Michigan, you can qualify to take the exam.

The journeyman exam is on the same subjects as the master exam but it covers only entry level knowledge. It will cost you $25 to apply for the journeyman exam. The license costs $20 and it's good for one year, expiring on December 31 each year.

Mechanical Contractor's License

To do mechanical work in Michigan you need a mechanical contractor's license. To get an application for a license, contact:

Contact

Michigan Department of Labor
Bureau of Construction Codes
Board of Mechanical Rules
7150 Harris Drive
P.O. Box 30255
Lansing, MI 48909
(517) 322-1798

The Board issues contractor licenses in these classifications:

Hydronic heating, cooling, process piping

HVAC equipment

Ductwork

Refrigeration

Limited heating service

Unlimited heating service

Limited refrigeration, air conditioning service

Unlimited refrigeration, air conditioning service

Fire suppression

Specialty

There are six types of specialty licenses: solar, solid fuel, LP tank and pipe, underground tank and pipe, gas piping, and gas piping and venting.

You will have to pass an exam to get a license. You need at least three years of experience or Board-approved equivalent in one or more of these classifications to qualify to take the exam. If you have Board-approved education you can apply up to one year of that education to the three-year work experience requirement.

Also if you want to take any of the unlimited service exams, you must first pass the limited service exam for that trade.

The exam has 65 multiple choice questions. There are 15 questions on mechanical contractor laws and rules, construction laws and rules, and basic safety rules. The Board will send you reference study materials for these laws and rules with your application. The other 50 questions are on the trade you want to get a license in. You have to pass the laws and rules part and at least one trade part to get a license.

All of the exams use the 1993 *State Mechanical Code* as a reference. The Board will sell you this book for $27. The other references depend on the type of license you're applying for.

Mechanical contractor's license fees: It'll cost you $25 for an exam but you can take as many trade exams at one time as you want for that fee. A license will cost you $75. You'll also have to pay $50 to the Construction Lien Recovery Fund to get licensed after you pass the exams. A license is good for one year, but all expire December 31 each year.

If you can prove you have a valid license in a state with qualifications at least equal to Michigan's (and your home state extends reciprocal licenses to Michigan mechanical contractors), you may be eligible for a Michigan license without taking the exam. You'll still need to pay the fee.

Recommended Reading for the Mechanical Trade Exams

Hydronic heating, cooling, process piping:

📖 *IBR Standard 200, Installation Guide for Residential Hydronic Heating Systems*, Hydronics Institute, 35 Russo Place, Berkeley Heights, NJ 07922

HVAC equipment:

📖 *HVAC Systems - Testing, Adjusting and Balancing*, 1993, SMACNA, 3221 W. Big Beaver, Suite #305, Troy, MI 48084

📖 *Manual J - Load Calculation for Residential Winter and Summer Air Conditioning*, 1986, 7th edition, Air Conditioning Contractors of America, 1712 New Hampshire Ave., NW, Washington, DC 20009

Ductwork:

📖 *Manual Q- Equipment Selection System Design Procedures for Commercial Summer and Winter Air Conditioning*, 1993, Air Conditioning Contractors of America, 1712 New Hampshire Ave., NW, Washington, DC 20009

📖 *Manual D- Duct Design Procedures for Residential Winter and Summer Air Conditioning and Equipment Selection*, 1993, Air Conditioning Contractors of America, 1712 New Hampshire Ave., NW, Washington, DC 20009

Hydronic heating, cooling, process piping:

📖 *Manual N - Load Calculation for Commercial Summer and Winter Air Conditioning*, 1988, 4th edition, Air Conditioning Contractors of America, 1712 New Hampshire Ave., NW, Washington, DC 20009

📖 *HVAC Systems - Testing, Adjusting and Balancing*, 1993, SMACNA, 3221 W. Big Beaver, Suite #305, Troy, MI 48084

Recommended Reading for the Mechanical Trade Exams (continued)

Refrigeration:

📖 *ASHRAE Standard 15-1992 Safety Code for Mechanical Refrigeration*, 1791 Tullie Circle NE, Atlanta, GA 30329

Fire suppression:

📖 *NFPA 13 - Installation of Sprinkler Systems*, 1985, National Fire Protection Association, 1 Batterymarch Park, P.O. Box 9101, Quincy, MA 02269

📖 *NFPA 13D- Installation of Sprinkler Systems in One and Two Family Dwellings and Mobile Homes*, 1991, National Fire Protection Association, 1 Batterymarch Park, P.O. Box 9101, Quincy, MA 02269

Solid fuel:

📖 *NFPA 211- Chimneys, Fireplaces, Vents and Solid Fuel Burning Appliances*, 1992, National Fire Protection Association, 1 Batterymarch Park, P.O. Box 9101, Quincy, MA 02269

Gas piping:

📖 *NFPA 54 - National Fuel Gas Code*, 1992, National Fire Protection Association, 11 Tracey Ave., Avon, MA 02322

Plumber's Licenses

To do plumbing work in Michigan you need a master or journeyman license. To get an application for a license, contact:

Contact

Michigan Department of Labor
Bureau of Construction Codes/Plumbing Board
7150 Harris Drive
P.O. Box 30254
Lansing, MI 48909
(517) 322-1804

You will have to pass an exam to get a license. To qualify for the master exam you need two years of work experience as a Michigan licensed journeyman plumber. You must have completed the experience just before you apply for the exam. If you got your experience in another state or in some other way, you can appear before the Board and present your case to take the exam. The written part of the exam is on the general theory and practice of plumbing and Michigan plumbing laws and regulations. The practical part of the exam will test your ability to do various mechanical tasks connected with the plumbing trade.

Fee

Plumber's license fees: It will cost you $50 to apply for the master exam. The license costs $75 and it's good for one year, expiring on December 31 each year. If you can prove you have a valid license in a state with qualifications at least equal to Michigan's, you may be eligible for a Michigan license without taking the exam. You still need to pay the fee though.

To qualify for the journeyman exam you must have at least three years experience as an apprentice plumber working under a master plumber. If you have Board-approved education you can use it for up to one year of the three years experience requirement. If you got your experience in another state or in some other way, you can appear before the Board and present your case to take the exam. The exam has a written part and a practical part. The written part lasts two and one-half hours and has 100 multiple choice questions. Here are the subjects on the written exam and the percentage of questions on each subject:

Subject	Percent of exam
Drainage systems	15
General knowledge	20
Fixtures	10
Water/backflow protection	20
Storm/special waste	5
Testing, inspections	5
Venting	15
Plumbing laws	10

The practical part of the exam lasts three hours and includes one full wiped solder joint (either straight or branch) and a copper project. The copper project is made up of a header system with one offset and two risers constructed from ¾- and ½-inch copper pipe with two threaded joints. The project must hold a pressure test and fit a template designed to measure the alignment and correct dimensions. The Board will supply the copper pipe, fittings, and solder. You have to provide soldering tools and the following:

- 1¾-inch x ¼-inch galvanized bushing
- 1¼-inch snifter valve
- joint sealing compound
- soldering flux

Fee

Journeyman's license fees: It will cost you $25 to apply for the journeyman exam. The license costs $20 and it's good for one year, expiring on December 31 each year. If you can prove you have a valid license in a state with qualifications at least equal to Michigan's, you may be eligible for a Michigan license without taking the exam. You still need to pay the fee though.

Department of Transportation (DOT)

You have to be prequalified to bid on Michigan Department of Transportation projects. To get an application, contact:

Contact

Michigan Department of Transportation
Financial Services Division
Prequalification Unit
P.O. Box 30050
Lansing, MI 48909
(517) 335-0137

Some of the details the Department will ask you about are your organization, personnel, financial condition, equipment, and experience. You'll be asked what type of work you want to be qualified for. Here are the types of work the Department uses:

Aggregate construction
Bridges, special structures
Building moving, demolition
Concrete curbs, gutters, driveways, sidewalks
Concrete pavement
Concrete pavement patching, widening
Electrical construction
Grading and drainage structures
Landscaping
Non-skid bituminous pavement
Plant-mixed bituminous pavement
Producing aggregate material
Pumphouses
Seeding and sodding
Structural steel
Structure concrete repair
Structural steel and prestressed concrete
Special contracts
Tunneling, jacking
Water mains, open cut sewers

Your prequalification rating is good for 15½ months after the end of the fiscal year your prequalification application is based on.

Out-of-State Corporations

An out-of-state corporation needs to register with the Michigan Department of Commerce to do business in the state. For information, contact:

Contact

Corporation Division
Michigan Department of Commerce
P.O. Box 30054
Lansing, MI 48909
(517) 334-6302
Fax: (517) 334-8329

Minnesota

The Minnesota Department of Commerce licenses residential builders and remodelers. Here's how you can contact them:

Contact

Department of Commerce
133 East 7th Street
St. Paul, MN 55101
(612) 296-6319
(800) 657-3978
(800) 657-3602 / (800) 925-5668 Minnesota only
Fax: (612) 296-8591

If you want to build or remodel residential property in Minnesota and your company's gross receipts are over $15,000 in a calendar year, you must get a license. If you don't expect to gross more than $15,000, you must still file an affidavit with the Department. If you go over the limit you must get a license immediately.

You don't need a license from the Department if you work as a plumber, electrician, or mechanical contractor. Plumbers are licensed by the Minnesota Department of Health and electricians are licensed by the Board of Electricity. Mechanical contractors are licensed by cities or counties but not by the state.

If you work in only one of the following specialty trades (except roofing) you don't need a residential builder/remodeler license. However if you work in two or more specialty trades, you need a license. Here are the trades the Department uses:

Excavation

Masonry/concrete

Carpentry

Interior finishing

Exterior finishing

Drywall and plaster

Roofing

General installation specialties

If you work in the roofing business, your company must get a roofer license. You (or an employee you designate) must also pass the residential roofer trade exam to get a license.

You must have liability insurance for at least $100,000 per occurrence and at least $10,000 property damage coverage. You may need unemployment insurance and Minnesota workers' compensation insurance. The application will also ask you about your company's organization and management personnel.

You'll also have to pay a fee to the Contractor's Recovery Fund. The fee is based on your company's gross receipts:

Company gross receipts	Recovery fund fee
Less than $1,000,000	$100
Over $1,000,000 and less than $5,000,000	$150
Over $5,000,000	$200

You must pass a business and law exam and a trade exam to get a license. The exams are given by Block and Associates. Each exam lasts two hours and has 50 multiple choice questions.

The business and law exam is closed book. Here's a table of the subjects on the exam and the percentage of questions on each subject:

Subject	Percent of exam
Contract management	26
Project management	18
Financial management	8
Tax law	12
Insurance and bonding	6
Business organization	2
Personnel regulations	12
Lien laws	6
Licensing	10

The reference for the business and law exam is the *Minnesota Contractors Reference Manual*. You can buy this and/or practice exams from Block and Associates by contacting them at:

Contact

Block and Associates
2100 NW 53rd Avenue
Gainesville, FL 32653-2149
(800) 280-3926

Here's a summary of the subjects on the Residential builder trade exam and the percentage of questions on the subject on the exam:

Subject	Percent of exam
Excavation and site work	13
Rough carpentry	26
Interior and exterior finish	23
Concrete and rebar	11
Roofing	6
Masonry	6
Estimating	6
Plans and specifications	4
Associated trades	5

Here's the information on the Residential remodeler trade exam and the percentage of questions on each subject:

Subject	Percent of exam
Excavation and site work	10
Rough carpentry	31
Interior and exterior finish	30
Concrete and rebar	5
Roofing	5
Masonry	5
Estimating	6
Plans and specifications	3
Associated trades	5

Here are the subjects on the Residential roofer trade exam and the percentage of questions on each subject:

Subject	Percent of exam
Steep roofing	32
Roofing components	25
Surface preparation	15
Low-slope roofing	10
Plans, specifications, estimating	10
Safety and employee protection	8

Fee

Builder, Remodeler and Roofer license fees: It will cost you $53 to take the business and law exam or a trade exam. A license will cost you $75. All licenses expire on March 31 each year.

Currently the Department has no reciprocity agreements with other states for these licenses.

Recommended Reading for the Builder, Remodeler and Roofer Exams

You can use these books during the exam:

📖 *Uniform Building Code , Volumes I, II, and III*, 1994 edition, International Conference of Building Officials, 5360 South Workman Mill Road, Whittier, CA 90601

📖 *Dwelling Construction Under the Uniform Building Code* , 1994 edition, International Conference of Building Officials, 5360 South Workman Mill Road, Whittier, CA 90601

You may want to study these but can't use them at the Residential builder or remodeler exams:

📖 *Carpentry and Building Construction*, 1993, Feirer, Hutchings & Feirer, McGraw-Hill Inc., Box 543, Blacklick, OH 43004-0543

📖 *Modern Carpentry,* 1992, Willis H. Wagner, The Goodheart-Willcox Company, Inc., 123 West Taft Drive, South Holland, IL 60473

📖 *Code of Federal Regulations - Title 29, Part 1926 (OSHA)*, July 1995, Superintendent of Documents, U.S. Government Printing Office, Washington, DC 20402

📖 *Minnesota Occupational Health and Safety Standards: Chapter 5207 - Standards for Construction*, Minnesota Department of Labor and Industry

You may want to study these but can't use them at the roofer's exam:

📖 *NRCA Roofing and Waterproofing Manual,* 1996, 4th edition, National Roofing Contractor's Association, 10255 W. Higgins, Rd., Suite #600, Rosemont, IL 60018-5607

📖 *Roofing Construction & Estimating*, 1995, Daniel Atcheson, Craftsman Book Company, 6058 Corte del Cedro, P.O. Box 6500, Carlsbad, CA 92018

📖 *Code of Federal Regulations - Title 29, Part 1926 (OSHA)*, July 1995, Superintendent of Documents, U.S. Government Printing Office, Washington, DC 20402

Asbestos Abatement Licenses and Certificates

To do asbestos work in Minnesota your business must be licensed by the Minnesota Department of Health. All employees of an asbestos business must also be certified by the Department. To get an application, contact:

Minnesota Department of Health
Asbestos Unit
121 East Seventh Place
P.O. Box 64975
St. Paul, MN 55164-0975
(612) 215-0900 / Fax: (612) 215-0975

To get an abatement contractor license you must employ a Department-certified site supervisor who will be responsible for the asbestos work your company does. You will also need to provide workers' compensation insurance and a Minnesota Business Identification Number.

Asbestos abatment license fee: The license will cost you $100 nonrefundable and it's good for one year.

Any individual working in asbestos abatement must be certified as a worker, site supervisor, inspector, management planner, or project designer. You must have completed the initial Department-approved asbestos training course associated with your trade before you apply for a certificate. You also need the following Department-approved education and/or experience requirements:

Worker

1. 1,000 hours in the general commercial construction trades; or,

2. 18-month vocational training program in construction; or,

3. two years in an apprenticeship program

Site supervisor

1. 2,000 hours in asbestos or general commercial construction trades; or,

2. bachelor's degree and 500 hours in asbestos or general commercial construction trades; or,

3. master's degree in environmental health, industrial hygiene or safety; or,

4. completion of an apprenticeship program

Inspector

1. 500 hours in building inspection, asbestos work, industrial hygiene, or hazardous materials control; or,

2. completion of an apprenticeship program; or,

3. Minnesota building official license; or,

4. bachelor's degree and 40 hours on-site asbestos inspection experience with a Minnesota-certified asbestos inspector; or,

5. registration as a professional engineer, architect, certified industrial hygienist, or certified safety professional

Management planner

1. 1,000 hours in building inspection, asbestos work, industrial hygiene, or hazardous materials control; or,

2. Minnesota building official license; or,

3. bachelor's degree and 500 hours in building inspection, asbestos work, industrial hygiene, or hazardous materials control; or,

4. registration as a professional engineer, architect, certified industrial hygienist, or certified safety professional; or,

5. master's degree in environmental health, industrial hygiene or safety and 250 hours in building inspection, asbestos work, industrial hygiene, or hazardous materials control

Project designer

1. 4,000 hours asbestos work or asbestos management; or,

2. registration as a professional engineer, architect, certified industrial hygienist, or certified safety professional

Asbestos abatement certificate fees: It will cost you $50 for a worker or site supervisor certificate, and $100 for any other certificate. A certificate is good for one year.

Electrician's Licenses

To do electrical work in Minnesota you need to be licensed by the State Board of Electricity. To get an application, contact:

Minnesota State Board of Electricity
1821 University Avenue, Room S-128
St. Paul, MN 55104-2993
(612) 642-0800
Fax: (612) 642-0441

The Board issues two types of contractor licenses — electrical, and alarm and communication. It also issues eight types of electrician licenses. Let's look first at the contractor licenses. To get either contractor license you must post a contractor's bond for $5,000. You must also have general liability insurance for $100,000 per occurrence, $300,000 aggregate, and property damage insurance for $300,000 per occurrence, $300,000 aggregate. There are no experience requirements for contractor licenses.

You don't have to pass an exam to get an electrical contractor license but you must employ a licensed master electrician who will be responsible for all the electrical work your company does.

However, you or an employee you designate, must pass an exam to get an alarm and communication contractor's license. The exam is on the current *National Electrical Code* and Ohm's and Watts laws. It will cost you $35 to take this exam.

Either contractor license will cost from $25 to $200, depending on when you get it. The electrical contractor license is good for two years, expiring the end of February each even-numbered year. The alarm and communication contractor license is good for two years, expiring June 30 each odd-numbered year.

Now let's look at the electrician licenses. Here's a summary of the requirements for each type of license and the annual license fee:

License type	Months of Board-approved work experience required	Annual license fee
Class A master	60	$40
Class A journeyman	48	$15
Master elevator constructor	60	$40
Elevator constructor	36	$15
Class A installer	12	$15
Class B installer	12	$15
Lineman	48	$15
Maintenance	48	$15

Fee

Electrician's exam license fee: The exam you have to pass to get an electrician license will be on licensing laws, the current *National Electrical Code*, and electrical theory. It will cost you $35 to take an exam. A license is good for two years.

The Board has some arrangements for reciprocal Class A master or journeyman licenses. If you've had a valid license in North or South Dakota for one year or more (and you passed their exam with a score of at least 70%), you can get a reciprocal license in Minnesota. If you've had a valid electrical license in another state for at least one year which you got through a licensing procedure at least equal to Minnesota's, you can get a similar Minnesota license without taking an exam. You'll have to pay a registration fee for your reciprocal license.

Plumber's Licenses

To do plumbing work in any city in Minnesota with a population over 5,000, you must get a license from the Minnesota Department of Health. To get an application, contact:

Contact

Minnesota Department of Health
Plumbing Unit
121 East Seventh Place, Suite 220
P.O. Box 64975
St. Paul, MN 55164-0975
(612) 215-0836
Fax: (612) 215-0977

The Department issues master and journeyman plumber licenses. If this is the first time you're applying for a license, you must include three notarized references with your application.

You must pass an exam to get a license. To qualify for the master exam you must have at least five years of Department-approved plumbing experience and a Minnesota journeyman plumber license or a master plumber license in a state with requirements equal to those in Minnesota.

To qualify for the journeyman plumber exam you need four years of Department-approved plumbing experience. If you have a journeyman plumber license in another state you must have passed an exam and had four years of practical plumbing work experience to get the license.

The exams are based on the *Minnesota Plumbing Code* and *Minnesota Rules*, Chapter 4715. Part of the exams are open book from these references. You can get these books from:

Contact

Minnesota Bookstore
117 University Avenue
St. Paul, MN 55155
(612) 297-3000
(800) 657-3757
Fax: (612) 296-2265

Fee

Plumber's license fees: It costs you $50 nonrefundable to take an exam. The master license costs $120 and the journeyman license is $55. A license is good for one year but they all expire on December 31 each year.

If you've passed a licensing exam in North or South Dakota and don't live in Minnesota, you can get a Minnesota license by reciprocity. Contact the Department for a reciprocity application.

Department of Transportation (DOT)

You don't have to be prequalified to bid on Minnesota Department of Transportation projects. However you may have to provide a written statement showing your experience, amount of capital, and equipment you have available for a project before you can get the contract for the job. The Department address is:

Department of Transportation
396 John Ireland Blvd.
St. Paul, MN 55155
(612) 296-6256
Fax: (612) 296-8773

Proposed projects are listed in the *Construction Bulletin*. For subscription information, contact:

Construction Bulletin
9443 Science Center Drive
New Hope, MN 55428
(612) 537-7730
(800) 328-4827 ext. 1162

The Department also has a free Bulletin Board Service (BBS) for contractors. To get information on this, contact:

Department of Transportation
Mail Stop 292
395 John Ireland Blvd.
St. Paul, MN 55155
(612) 297-1237

You need to know about the Department's requirements for its construction projects before bidding on a project. You can get this information in their publication, *Standard Specifications for Construction*. The Department sells it for $20 hard cover or $17 soft cover plus tax, shipping, and handling. Here's the contact office:

Manual Sales Office
Department of Transportation
Mail Stop 260
395 John Ireland Blvd.
St. Paul, MN 55155
(612) 296-2216

Out-of-State Corporations

Out-of-state corporations should check with the Minnesota Secretary of State before doing business in Minnesota. You can contact them at:

Contact

Secretary of State
180 State Office Building
St. Paul, MN 55155
(612) 296-2803

Mississippi

Depending on the size and cost of a commercial construction or remodeling project you want to work on in Mississippi, you may need a Certificate of Responsibility from the Mississippi Contractor Licensing Board. To get particular regulations on size and cost or an application, contact:

Contact

Mississippi Contractor Licensing Board
2001 Airport Road, Suite 101
Jackson, MS 39208
(601) 354-6161
(800) 354-6161
Fax: (601) 354-6715

Commercial Construction Contractor's Certificate

The Board issues certificates in these classifications:

Building construction
Heavy construction
Highway, street, bridge construction
Mechanical work
Municipal, public works construction
Electrical work
Specialty work

On the application, the Board will ask you about your company organization, personnel, financial condition, equipment, and experience. You must have workers' compensation insurance and liability insurance for at least $300,000 per occurrence and $600,000 aggregate. You'll need federal and Mississippi tax numbers, and a Certificate of Authority from the Mississippi Secretary of State. To apply for them, contact:

Contact

Secretary of State
P.O. Box 136
Jackson, MS 39205
(601) 359-1633
Fax: (601) 359-1607

Commercial construction contractor's certificate fees: You've got to pass a business and law exam and a trade exam for the classification(s) you want to be certified in. Each exam costs $50. It will cost you $100 for a certificate in one classification. Additional classifications cost $50 each. A certificate is good for one year.

The exams are given by Block Testing Services. You can contact them at:

Block State Testing Services L.P.
2100 NW 53rd Avenue
Gainesville, FL 32653-2149
(800) 280-3926

The business and law exam has 50 multiple choice questions and lasts three hours. Here's a summary of the subjects on the exam and the approximate number of questions for each subject:

Subject	Number of questions
Contract management	10 - 12
Employment law and regulations	9 - 11
Project management	6 - 8
Payroll taxes	5 - 7
Licensing laws, regulations	4 - 6
Financial management	4 - 6
Insurance and bonding	2 - 4
Lien laws	1 - 3
Business organization	1 - 3

The reference for the exam is the *Mississippi Contractor's Reference Manual*. You can get it from Block for $55. They also sell a practice exam for $24.

Here's the same information for each trade classification, followed by a reference list for the exam:

Building Construction Certificate

The test is 50 multiple choice questions, three hours:

Subject	Number of questions
Concrete materials	2 - 4
Concrete methods, installation	4 - 6
Reinforcing steel	2 - 4
Structural steel	7 - 9
Wood design	7 - 9

Subject	Number of questions
Safety, OHSA	7 - 9
Formwork	5 - 7
Masonry	4 - 6
Drywall installation	3 - 5

Heavy Construction Certificate

The test is 80 multiple choice questions, four hours:

Subject	Number of questions
Concrete	4 - 8
Safety, Code of Federal Regulations	1 - 3
Concrete, reinforcing steel	1 - 3
Cut sheet calculations	3 -5
Hydraulic dredging	3 - 5
Pile driving	3 - 5
Sanitary/storm sewers	3 - 5
General knowledge	1 - 3
Ductile pipe installations	2 - 4
Concrete pipe installations	2 - 4
Clay pipe installations	2 -4
Blasting	4 - 6
General plant construction	4 - 6
Asphalt paving procedures	1 - 3
Asphalt properties, general uses	1 - 3
Asphalt surface treatments	1 - 3
Excavating soils	4 - 8
Excavating, grading	4 - 8
Structural steel	4 - 8
Fuel gas systems, pressure gas pipelines	1 - 3
Rigging	4 - 8

Highway Construction Certificate

The test is 50 multiple choice questions, three hours:

Subject	Number of questions
Concrete	4 - 8
Structural steel	4 - 8
Concrete, reinforcing steel	4 - 8
Pile driving	1 - 4

Subject	Number of questions
Sanitary/storm sewers	4 - 8
Rigging	4 - 8
Asphalt paving procedures	1 - 4
Asphalt properties, general uses	1 - 4
Asphalt surface treatments	1 - 4
Excavating soils	4 - 8
Excavating, grading	5 - 10

Electrical Work Certificate

This is a six-hour exam in three parts:

Part I — *closed book, one hour, 50 multiple choice questions, 21% of total score*

Part II — *open book, two hours, 70 multiple choice questions, 29% of total score*

Part III — *open book, three hours, 30 multiple choice questions, 50% of total score*

Subject	Number of questions
Calculations	40 - 50
General use equipment	30 - 40
Electrical theory	18 - 24
Wiring methods, materials	12 - 18
Special equipment	6 - 8
Wiring and protection	6 - 8
Special occupancies	5 - 7
Low voltage systems	3 - 5

Mechanical Work Certificate

The test is 80 multiple choice questions, four hours:

Subject	Number of questions
Code compliance	5 - 10
Fuel gas systems	4 - 6
Safety	4 - 8
HARV — general	6 - 12
HARV — maintenance	6 - 12
HARV — controls	5 - 10
Boilers — high pressure	3 - 5
Boilers — low pressure	3 - 5
Boilers — safety	1 - 3

Subject	Number of questions
Ducts	5 - 10
Fan theory	1 - 3
Ducts — safety, fire prevention	1 - 3
Ducts — vapor removal, hood systems	1 - 3
Piping	5 - 10
Insulation	1 - 3
Psychrometric analysis	2 - 4
HARV — load calculation	4 - 8

Recommended Reading for the Commercial Construction Contractor's Certificates

Building Construction Certificate

- *Code of Federal Regulations - Title 29,* Part 1926 (OSHA), July 1995, Superintendent of Documents, U.S. Government Printing Office, Washington, DC 20402

- *Design and Control of Concrete Mixtures,* 1991, Kosmatka and Panarese, Portland Cement Association, 5420 Old Orchard Road, Skokie, IL 60077-1083

- *Concrete Masonry Handbook,* 1991, 5th edition, Portland Cement Association, 5420 Old Orchard Road, Skokie, IL 60077-1083

- *Handling and Erection of Steel Joists and Joist Girders, Technical Digest No. 9,* 1987, Steel Joist Institute, 1205 48th Avenue, North - Suite A, Myrtle Beach, SC 29577

- *Code of Standard Practice for Steel Buildings and Bridges,* 1992, American Institute of Steel Construction, P.O. Box 806276, Publication Department, Chicago, IL 60680-4124

- *Specification for Structural Joints Using ASTM A325 or A490 Bolts,* ASD, 1986, American Institute of Steel Construction, P.O. Box 806276, Publication

- *Span Tables for Joists and Rafters,* 1993 American Forest and Paper Association, 1111 19th Street NW, Washington, DC 20002

- *Recommended Specifications for the Application and Finishing of Gypsum Board-GA-216-93,* 1993, Gypsum Association, 810 First Street NE, Washington, DC 20002

- *ASTM C-94 Standard Specification for Ready Mixed Concrete,* 1994, ASTM, 100 Bar Harbour Drive, West Conshohocken, PA 19428-2959

- *Commentary and Recommendations for Handling, Installing and Bracing Metal Plate Connected Wood Trusses,* HIB-91, Truss Plate Institute, 583 D'Onofrio Drive, Madison, WI 53719

Recommended Reading for the Commercial Construction Contractor's Certificates (continued)

Heavy Construction Certificate

- *Code of Federal Regulations - Title 29,* Part 1926 (OSHA), July 1995, Superintendent of Documents, U.S. Government Printing Office, Washington, DC 20402

- *Design and Control of Concrete Mixtures,* 1991, Kosmatka and Panarese, Portland Cement Association, 5420 Old Orchard Road, Skokie, IL 60077-1083

- *Cement Mason's Guide,* 1990, Portland Cement Association, 5420 Old Orchard Road, Skokie, IL 60077-1083

- *Construction Planning, Equipment and Methods,* 1995, 5th edition, R. L. Peurifoy, McGraw-Hill Publishing, Blue Ridge Summit, PA 17294

- *Design of Municipal Wastewater Treatment Plants,* 1992, Water Environment Federation, 601 Wythe St., Alexandria, VA 22314-1994

- *Hydraulic Dredging: Principles - Equipment - Procedures - Methods,* 1986, John Huston, C.E., P.O. Box 6372, Corpus Christi, TX 78411

- *ASME B31.8 - Gas Transmission and Distribution Piping Systems,* 1995, American Society of Mechanical Engineers, 22 Law Drive, P.O. Box 2300, Fairfield, NJ 07007-2300

- *Handbook of Rigging,* 1988 4th edition, W. E. Rossnagel, McGraw-Hill Publishing, Inc., Blue Ridge Summit, PA 17294

- *ASTM C-94 Standard Specification for Ready Mixed Concrete,* 1994, ASTM, 100 Bar Harbour Drive, West Conshohocken, PA 19428-2959

- *Clay Pipe Engineering Manual,* 1995, National Clay Pipe Institute, P.O. Box 759, Lake Geneva, WI 53147

- *Clay Pipe Installation Handbook,* 1994, National Clay Pipe Institute, P.O. Box 759, Lake Geneva, WI 53147

- *AWWA Standard for Installation of Ductile Iron Water Mains,* AWWA C-600-93, 1993, American Water Works Association, 6666 W. Quincy Ave., Denver, CO 80235

- *Concrete Pipe Installation Manual,* 1995, American Concrete Pipe Association, 222 West Las Colinas Blvd., Suite 641, Irving, TX 75039-5423

- *MS-5 - Introduction to Asphalt,* 1986, 8th edition, Asphalt Institute, P.O. Box 14052, Lexington, KY 40512-4052

- *MS-11 - Asphalt Surface Treatments - Specifications,* 1991, Asphalt Institute, P.O. Box 14052, Lexington, KY 40512-4052

Recommended Reading for the Commercial Construction Contractor's Certificates (continued)

Heavy Construction Certificate (continued)

📖 *MS-12- Asphalt Surface Treatments - Construction Techniques,* 1988, Asphalt Institute, P.O. Box 14052, Lexington, KY 40512-4052

📖 *Excavation & Grading Handbook,* 1991, Nicholas E. Capachi, Craftsman Book Company, 6058 Corte del Cedro, P.O. Box 6500, Carlsbad, CA 92018

📖 *Manual of Steel Construction,* 1989, 9th edition, American Institute of Steel Construction, P.O. Box 806276, Chicago, IL 60680-4124

Highway Construction Certificate

📖 *Code of Federal Regulations - Title 29, Part 1926 (OSHA),* July 1995, Superintendent of Documents, U.S. Government Printing Office, Washington, DC 20402

📖 *Design and Control of Concrete Mixtures,* 1991, Kosmatka and Panarese, Portland Cement Association, 5420 Old Orchard Road, Skokie, IL 60077-1083

📖 *Construction Planning, Equipment and Methods,* 1995, 5th edition, R. L. Peurifoy, McGraw-Hill Publishing, Blue Ridge Summit, PA 17294

📖 *Handbook of Rigging,* 1988 4th edition, W. E. Rossnagel, McGraw-Hill Publishing, Inc., Blue Ridge Summit, PA 17294

📖 *MS-5 - Introduction to Asphalt,* 1986, 8th edition, Asphalt Institute, P.O. Box 14052, Lexington, KY 40512-4052

📖 *MS-11- Asphalt Surface Treatments - Specifications,* 1991, Asphalt Institute, P.O. Box 14052, Lexington, KY 40512-4052

📖 *MS-12- Asphalt Surface Treatments - Construction Techniques,* 1988, Asphalt Institute, P.O. Box 14052, Lexington, KY 40512-4052

📖 *Excavation & Grading Handbook,* 1991, Nicholas E. Capachi, Craftsman Book Company, 6058 Corte del Cedro, P.O. Box 6500, Carlsbad, CA 92018

📖 *Manual of Steel Construction,* 1989, 9th edition, American Institute of Steel Construction, P.O. Box 806276, Chicago, IL 60680-4124

Electrical Work Certificate

📖 *NFPA 70 - National Electrical Code,* 1996 edition, National Fire Protection Association, 1 Batterymarch Park, P.O. Box 9101, Quincy, MA 02269

📖 *American Electricians Handbook,* 1996, 13th edition, Croft/Summers, McGraw-Hill Inc., Box 543, Blacklick, OH 43004-0543

Recommended Reading for the Commercial Construction Contractor's Certificates (continued)

Electrical Work Certificate (continued)

📖 *Electrical Review for Electricians,* 9th edition, 1996, J. Morris Trimmer and Charles Pardue, Construction Bookstore, 100 Enterprise Place, Dover, DE 19903-7029

📖 *National Electrical Code Blueprint Reading,* K. L. Gebert, American Technical Publishers, 1155 West 175th Street, Homewood, IL 60403

Mechanical Work Certificate

📖 *Refrigeration and Air Conditioning,* 1995, 3rd edition, Air Conditioning and Refrigeration Institute, Prentice Hall, P.O. Box 11071, Des Moines, IA 50336-1071

📖 *Standard Gas Code 1994,* Southern Building Code Congress International, Inc., 900 Montclair Road, Birmingham, AL 35213-1206

📖 *Standard Mechanical Code 1994,* Southern Building Code Congress International, Inc., 900 Montclair Road, Birmingham, AL 35213-1206

📖 *Trane Ductulator,* 1976, Trane Company, 8929 Western Way, Suite #1, Jacksonville, FL 32256

📖 *Manual D- Duct Design Procedures for Residential Winter and Summer Air Conditioning and Equipment Selection,* 1993, Air Conditioning Contractors of America, 1712 New Hampshire Ave., NW, Washington, DC 20009

📖 *Manual N - Load Calculation for Commercial Summer and Winter Air Conditioning,* 1988, 4th edition, Air Conditioning Contractors of America, 1712 New Hampshire Ave., NW, Washington, DC 20009

📖 *Low Pressure Boilers,* 1994, F. M. Steingress, American Technical Publishers, 1155 West 175th Street, Homewood, IL 60403

📖 *High Pressure Boilers,* 1993, F. M. Steingress, American Technical Publishers, 1155 West 175th Street, Homewood, IL 60403

📖 *Pipefitters Handbook,* 1967, 3rd edition, Forest Lindsey, Industrial Press, Inc., 200 Madison Avenue, New York, NY 10016

📖 *Code of Federal Regulations - Title 29, Part 1926 (OSHA),* July 1995, Superintendent of Documents, U.S. Government Printing Office, Washington, DC 20402

📖 *ANSI/ASHRAE Standard 15-94 Safety Code for Mechanical Refrigeration,* American National Standards Institute, 11 West 42nd Street, New York, NY 10036

Recommended Reading for the Commercial Construction Contractor's Certificates (continued)

Mechanical Work Certificate (continued)

📖 *Single Burner Boiler Operation,* National Fire Protection Association, 1 Batterymarch Park, Box 9101, Quincy, MA 02269-9101

📖 *NFPA 90A - Installation of Air Conditioning and Ventilating Systems,* 1993, National Fire Protection Association, 1 Batterymarch Park, Box 9101, Quincy, MA 02269-9101

📖 *NFPA 96 - Standard for Ventilation Control and Fire Protection of Commercial Cooking Operations,* 1994, National Fire Protection Association, 1 Batterymarch Park, Box 9101, Quincy, MA 02269-9101

Municipal Construction Certificate

The test is 50 multiple choice questions, three hours:

Subject	Number of questions
General plant construction	8 - 12
Sanitary, storm sewers	4 - 8
Excavating, grading	6 - 10
Ductile pipe installation	2 - 5
Concrete pipe installation	2 - 5
Clay pipe installation	2 - 5
Concrete reinforcing steel	3 - 7
Cut sheet calculations	3 - 5
Safety, OSHA	1 - 3

The Board reviews applications from out-of-state for reciprocity.

Recommended Reading for the Municipal Construction Certificate

📖 *Code of Federal Regulations - Title 29, Part 1926 (OSHA),* July 1995, Superintendent of Documents, U.S. Government Printing Office, Washington, DC 20402

📖 *Design and Control of Concrete Mixtures,* 1991, Kosmatka and Panarese, Portland Cement Association, 5420 Old Orchard Road, Skokie, IL 60077-1083

📖 *Construction Planning, Equipment and Methods,* 1995, 5th edition, R. L. Peurifoy, McGraw-Hill Publishing, Blue Ridge Summit, PA 17294

📖 *Design of Municipal Wastewater Treatment Plants,* 1992, Water Environment Federation, 601 Wythe St., Alexandria, VA 22314-1994

**Recommended Reading for the
Municipal Construction Certificate (continued)**

📖 *ASTM C-94 Standard Specification for Ready Mixed Concrete*, 1994, ASTM, 100 Bar Harbour Drive, West Conshohocken, PA 19428-2959

📖 *Clay Pipe Engineering Manual*, 1995, National Clay Pipe Institute, P.O. Box 759, Lake Geneva, WI 53147

📖 *Clay Pipe Installation Handbook*, 1994, National Clay Pipe Institute, P.O. Box 759, Lake Geneva, WI 53147

📖 *AWWA Standard for Installation of Ductile Iron Water Mains*, AWWA C-600-93, 1993, American Water Works Association, 6666 W. Quincy Ave., Denver, CO 80235

📖 *A Guide for the Installation of Ductile Iron Pipe*, 1988, Ductile Iron Pipe Research Association, 245 Riverchase Parkway, East, Suite O, Birmingham, AL 35244

📖 *Concrete Pipe Installation Manual*, 1995, American Concrete Pipe Association, 8300 Boone Blvd., Suite 400, Vienna, VA 22182

📖 *Excavation & Grading Handbook*, 1991, Nicholas E. Capachi, Craftsman Book Company, 6058 Corte del Cedro, P.O. Box 6500, Carlsbad, CA 92018

You can get the references for any of the exams from:

Builder's Book Depot
1033 E. Jefferson St., Suite 500
Phoenix, AZ 85034
(800) 284-3434

Residential Building and Remodeler's Licenses

You must have a Residential Building license to build any residence which is three stories or less in height if the cost of the project is more than $50,000. You must have a Residential Remodeler's license to remodel any residence if the cost of the project is more than $10,000. If you don't reside in Mississippi and you have a license from a state whose licensing requirements are equal to those in Mississippi, you don't need a Mississippi license. To get an application for a residential license, contact:

Contact

Mississippi Contractor Licensing Board
2001 Airport Road, Suite 101
Jackson, MS 39208
(601) 354-6161 / (800) 354-6161
Fax: (601) 354-6715

On the application the Board will ask you about your company organization, personnel, financial condition, equipment, and experience. You must have workers' compensation insurance and liability insurance. You will need federal and Mississippi tax numbers. You also need a Certificate of Authority from the Mississippi Secretary of State. To apply for this contact:

Secretary of State
P.O. Box 136
Jackson, MS 39205
(601) 359-1633
Fax: (601) 359-1607

You must pass a Residential trade exam to get a license. The exam is given by Block Testing Services. You can contact them at:

Block State Testing Services L.P.
2100 NW 53rd Avenue
Gainesville, FL 32653-2149
(800) 280-3926

The Residential exam lasts three hours and has 50 multiple choice questions. Here are the subjects on the exam and the approximate number of questions for each subject:

Subject	Number of questions
Business	14 - 16
Concrete	4 - 6
Wood	7 - 9
Foundations	2 - 4
Drywall	4 - 6
Roofing	3 - 5
Associated trades	2 - 4
Masonry	3 - 5
Safety	2 - 4

Residential building license fees: It will cost you $50 to file an application for a license. The exam costs $50. A license is good for one year.

The Board has reciprocity agreements with Tennessee, Arkansas, Alabama, and Louisiana.

Recommended Reading for the Residential Building Exam

📖 *Mississippi Contractor's Reference Manual*, Block & Associates, 2100 NW 53rd Avenue, Gainesville, FL 32653

📖 *Standard Building Code 1994*, Southern Building Code Congress International, Inc., 900 Montclair Road, Birmingham, AL 35213-1206

Recommended Reading for the Residential Building Exam (continued)

📖 *Carpentry and Building Construction*, 1993, Feirer, Hutchings & Feirer, McGraw-Hill Inc., Box 543, Blacklick, OH 43004-0543

You can get any of these references from:

Builder's Book Depot
1033 E. Jefferson St., Suite 500
Phoenix, AZ 85034
(800) 284-3434

Department of Transportation (DOT)

You have to be bonded to bid on Mississippi Department of Transportation projects. You also have to appoint a Mississippi resident to be your agent in the state. Depending on the project, you may have to get a Certificate of Responsibility from the State Board of Contractors. For information on this, contact:

Department of Transportation
Contract Administration Division
P.O. Box 1850
Jackson, MS 39215-1850
(601) 359-7700
Fax: (601) 359-7732

Out-of-State Corporations

Out-of-state corporations should check with the Mississippi Secretary of State before doing business in Mississippi. You can contact them at:

Secretary of State
P.O. Box 136
Jackson, MS 39205
(601) 359-1333
Fax: (601) 359-1607

Missouri

The state of Missouri doesn't license construction contractors.

Department of Transportation (DOT)

To bid on Missouri Department of Transportation projects costing less than $2,000,000, you need to fill in the Contractor Questionnaire form from the Department. If you want to bid on projects costing $2,000,000 or more, you need to fill in the Prequalification Contractor Questionnaire form. To get these forms, contact:

Contact

Missouri Department of Transportation
105 West Capitol Avenue
P.O. Box 270
Jefferson City, MO 65102
(573) 751-2551
Fax: (573) 751-6555

Some of the things the Contractor Questionnaire form will ask you are:

- who the officers of your company are
- what construction equipment you own
- what your work experience is
- what work you've completed in the last three years
- what the financial condition of your company is

The Prequalification Contractor Questionnaire form will also ask you what classes of work you want to be prequalified in. Here are the classes of work the Department uses:

Earthwork

Bituminous pavement

Bridges, culverts, etc.

Portland cement concrete pavement

Other

You'll also be asked to designate a Missouri resident who will be your agent for receipt of legal process. Prequalification is good for up to 24 months.

Out-of-State Corporations

You must register with the Missouri Secretary of State and get a Certificate of Good Standing to do business in the state. To get an application contact:

Contact

Business Services
P.O. Box 778
Jefferson City, MO 65102
(573) 751-4153
Fax: (573) 751-5841

Montana

In Montana, all construction contractors and subcontractors must register with the Department of Labor and Industry. To get an application to register, contact:

Contact

Contractors Registration Unit
Employment Relations Division
Department of Labor and Industry
P.O. Box 8011
Helena, MT 59604-8011
(406) 444-7734
Fax: (406) 444-4140

If you're a licensed electrician or plumber and work only in those trades, you don't need to register with the Department. You do need a Montana license, though. See below for more information on these trades.

Fee

Contractor's registration fees: You won't have to take an exam to register with the Department. Registration costs $70 and it's good for two years.

Electrician's Licenses

To do electrical work in Montana, you must get a license. Montana licenses electrical contractors, master, journeyman, and residential electricians. To apply for a license, contact:

Contact

Montana State Electrical Board
111 North Jackson
P.O. Box 200513
Helena, MT 59620-0513
(406) 444-4390
Fax: (406) 444-1667

Electrical Contractor's License

To qualify for an electrical contractor's license, you must have a master electrician's license you got by testing in Montana or employ a licensed master electrician who will be responsible for the electrical work you do. You don't have to pass an exam to get a contractor license. The license costs $250 and it's good for three years.

You will have to pass an exam to get a master, journeyman, or residential electrician's license. The exam is given by National Assessment Institute (NAI). For information on the exam, contact:

Contact

National Assessment Institute, Inc.
560 East 200 South, Suite 300
Salt Lake City, UT 84102
(801) 355-5009

Master Electrician's License

To apply for the master electrician's exam you must prove you can meet one of these three requirements:

- an electrical engineering degree and one year practical electrical work experience
- graduation from a post-secondary electrical trade school and four years practical work experience
- five years of practical work experience with wiring, apparatus, or equipment for electrical light, heat and power

The master electrician's exam has at least 30 questions covering these subjects:

- *National Electrical Code*
- cost estimating for electrical installments
- procurement and handling electrical installation and repair materials
- reading electrical blueprints
- drafting and laying out electrical circuits
- practical electrical theory knowledge

Journeyman Electrician's License

To apply for a journeyman electrician's license you must prove you can meet either of these two requirements:

- completion of an apprenticeship program of four years (8,000 hours) electrical work
- four years (8,000 hours) of practical work experience with wiring, apparatus, or equipment for electrical light, heat and power

No more than 4,000 hours can be in residential construction. Remaining hours must be in commercial, industrial, or institutional construction. The Board may give you credit for hours you spend in a recognized electrical trade school.

The journeyman electrician's exam has at least 30 questions covering these subjects:

- *National Electrical Code*
- Ohm's law
- laying out practical electrical circuits

Residential Electrician's License

To apply for a residential electrician's license you must prove you can meet either of these two requirements:

- completion of a two-year electrical apprenticeship
- two years of practical work experience with wiring, apparatus, or equipment for electrical light, heat and power in residential construction

The residential electrician's exam has at least 30 questions covering these subjects:

- *National Electrical Code*
- Ohm's law
- laying out residential electrical circuits

Fee

Electrician's license fees: It will cost you $20 nonrefundable to file an application for a license. An exam will cost you $50. The license will cost $100 and it's good for three years.

If you can prove you have a valid license in a state with qualifications at least equal to Montana's and you have the work experience Montana requires, you can get a Montana journeyman license without taking an exam. You'll still need to pay a $145 fee for the license.

Plumber's Licenses

To do plumbing work in Montana you must get a license. To apply for a license, contact:

Contact

Montana State Board of Plumbers
111 North Jackson
P.O. Box 200513
Helena, MT 59620-0513
(406) 444-4390
Fax: (406) 444-1667

The Board licenses master and journeyman plumbers. You must pass an exam given by the Board to get a license.

Master Plumber's License

To qualify for a master plumber's license you must prove you meet these requirements:

- four years (1,500 hours per year) work experience as a journeyman plumber
- three years work experience with a licensed master plumber or as a supervisor in plumbing work

Journeyman Plumber's License

To apply for a journeyman plumber's license you must prove you have five years of practical plumbing work experience or have completed a Board-approved apprenticeship program.

Plumber's license fees: It will cost you $30 nonrefundable to file an application for a license and $95 to take an exam. A master license costs $125 and a journeyman license costs $75. Licenses expire August 31 each year.

If you can prove you have a valid license in a state with qualifications at least equal to Montana's and you have the work experience Montana requires, you can get a Montana journeyman license without taking an exam. You still need to pay a $95 endorsement fee for the license.

Department of Transportation (DOT)

The Montana Department of Transportation doesn't prequalify contractors. To bid on highway construction projects in Montana you must get a license through the Montana Department of Labor. See the beginning of this section for more information.

Out-of-State Corporations

Out-of-state corporations must register with the Montana Secretary of State to do business in the state. For information contact:

Business Services Bureau
State Capitol, Room 225
Helena, MT 59620
(406) 444-2034
Fax: (406) 444-3976

Nebraska

To do business as a contractor in Nebraska, you don't need a license if you're a resident. But as a nonresident contractor in Nebraska, you must register with the Nebraska Secretary of State and the Nebraska Department of Revenue. You'll find information on the Secretary of State under the heading Out-of-State Corporations, at the end of this chapter. To register with the Department of Revenue, contact:

Contact

Nebraska Department of Revenue
301 Centennial Mall South
P.O. Box 94818
Lincoln, NE 68509-4818
(402) 471-2971
Fax: (402) 471-5608

It'll cost you $25 to register with the Department.

You must post a bond for any contract you get in the state for more than $2,500. The bond is $1,000 for a contract between $2,500 and $10,000. For a contract over $10,000, the bond must be 10% of the contract amount for the first $100,000 and 5% of any amount over $100,000.

Electrician's Licenses

To do electrical work in Nebraska you need to be licensed by the Nebraska State Electrical Division. Some cities in the state also require you to have a license. To get an application, contact:

Contact

State of Nebraska
State Electrical Division
800 South 13th Street, Suite 109
P.O. Box 95066
Lincoln, NE 68509-5066
(402) 471-3550
Fax: (402) 471-4297

The Division issues contractor, journeyman, and fire alarm installer licenses. You have to pass an exam to get your license.

To qualify for the contractor's exam, you need one of the following:

- degree from an accredited college or university in a four-year electrical course
- one year Division-approved work experience as a journeyman electrician
- five years Division-approved work experience

To get a contractor's license you must also have liability insurance for $100,000 for each person, $300,000 for each accident, and $100,000 property damage.

For the journeyman exam you need four years Division-approved work experience or three years Division-approved work experience and a course in electrical wiring from a Division-approved school. For the fire alarm installer exam you need two years Division-approved work experience.

The exam will be open book, based on the *National Electrical Code*. It has at least 50 questions and lasts three hours. There will also be questions on basic electricity, blueprint reading, emergency circuits, and the Nebraska State Electrical Act.

Here's a summary of the exam and license fees for each license. The license fee varies depending on whether you get it in an even- or odd-numbered year.

License	Exam fee	License fee Even-numbered year	Odd-numbered year
Contractor	$62.50	$75	$150
Journeyman	$12.50	$15	$30
Fire alarm installer	$12.50	$15	$30

Licenses expire on December 31 in even-numbered years.

If you've had a valid South Dakota license for at least one year and you got it by passing an exam, you can apply for a reciprocal Nebraska license. You must also never have failed a Nebraska state electrical licensing exam. You still need to pay the Nebraska exam and license fees to get a reciprocal license.

You can also get a reciprocal state license if you have a valid city license in Fremont, Hastings, Lincoln, or Omaha, Nebraska. You must have passed the city exam within the last three years before you apply for a state license. You'll also have to pay the state license fee.

Department of Roads

To bid on Nebraska Department of Roads projects, you have to be prequalified by the Department. To get an application for prequalification, contact:

Nebraska Department of Roads
Contract Lettings Division
1500 Nebraska Highway 2
P.O. Box 94759
Lincoln, NE 68509-4759
(402) 479-4525
Fax: (402) 479-4325

Some of the details the Department will ask you about are your organization, personnel, financial condition, equipment, and experience. You'll be asked what type of work you want to be qualified for. Here are the major types of work the Department uses:

Grading	Guardrails, fencing
Aggregates	Landscaping
Bituminous	Bridges
Concrete pavement	General
Culverts	

The Department also uses these specialty construction types of work:

Building construction	Painting
Electrical	Demolition
Signing	

Prequalification is good for 15 months.

Out-of-State Corporations

Out-of-state corporations must get a Certificate of Authority to do Business in Nebraska from the Nebraska Secretary of State. To apply for this certificate, contact:

Nebraska Secretary of State
Corporations Division
State Capitol 1301
Lincoln, NE 68509
(402) 471-2554
Fax: (402) 471-3666

Nevada

You must be licensed to bid or work on construction jobs in Nevada. To apply for a license, ask for a license packet from:

Contact

State Contractors Board
4220 S. Maryland Parkway, Suite D-800
Las Vegas, NV 89119
(702) 487-1100
Fax: (702) 486-1190

The Board has these primary classifications for contractors:

General engineering

General engineering and general building

General building

Plumbing and heating

Electrical

Carpentry

Painting and decorating

Concrete

Erecting signs

Elevation and conveyance

Glass and glazing

Movement of buildings

Landscape contracting

Spraying mixtures containing cement

Using sheet metal

Steel reinforcing and erection

Roofing and siding

Finishing floors

Lathing and plastering

Masonry

Tiling

Refrigeration and air conditioning

Drilling wells and installing pumps

Erecting scaffolds and bleachers

Fencing and equipping playgrounds

Institutional contracting

Individual sewerage

Fabricating tanks

Installing equipment to treat water

Wrecking

Setting refractories

Installing industrial machinery

Installing bowling alleys

Installing vaults and safes

Installing urethane

Solar contracting

Installing equipment used with liquefied petroleum and natural gas

Installing heaters

Fire protection

Constructing, altering, or improving community antenna television systems

Other

Most classifications also have subclassifications. You need to give the Board four notarized references verifying that you've had four years of work experience in a classification to apply for a license for it.

When the Board gets your completed application, they'll check it and decide if you're eligible to take the exams they require for licensing. If so, they'll send you an exam registration form to fill in. You must take an exam called the Contractors Management Survey Exam (CMS). It's given once each month by the National Assessment Institute (NAI). For information on the exam contact:

Contact

National Assessment Institute, Inc.
560 East 200 South, Suite 300
Salt Lake City, UT 84102
(801) 355-5009

The CMS exam is on:

- Nevada State Contractor licensing laws and regulations
- state laws
- construction project management
- business and financial management

It's open book and lasts three hours.

To get a license in most of the classifications you also need to pass an exam in that classification. Trade classification exams are on:

- reading and interpreting construction codes and regulations
- Nevada Health Department regulations
- building codes
- trade materials, tools, equipment, and methods
- Nevada Occupational Safety and Health rules

A trade exam is given in two sessions — a three-hour morning session and an afternoon session for three or four hours.

You can get a reading list for the trade part of the exam from the State Contractors Board. The list will tell you which code books, state laws and regulations, technical manuals, and textbooks you need.

Nevada may waive the trade exam if you can verify that you've been actively licensed in that trade in Arizona, California, or Utah for five of the last seven years. However, if you're applying for a plumbing or electrical trade license, you must take the trade exam.

Fee

Contractor's license fees: If you want to be licensed in more than one trade you have to apply and pay for it separately. Each license will cost you $185 nonrefundable for the application, $150 for the license, $40 for the CMS exam, and $35 for a trade exam.

You will also need to get a surety bond. The Board sets the amount of the bond. You must also prove you have industrial insurance with the Nevada State Industrial Insurance System. In Las Vegas their phone number is (702) 388-3114.

Recommended Reading for the CMS Exam

The exam administrators will provide these books for you to use at the exam:

Nevada Contractor's Reference Manual

National Electric Code

Uniform Building Code

Uniform Plumbing Code

Uniform Mechanical Code

Uniform Fire Code

NAI also recommends *Builder's Guide to Accounting* by Michael Thomsett (from Craftsman Book Company, P.O. Box 6500, Carlsbad, CA 92018) as a reference for the CMS exam.

Out-of-State Corporations

Out-of-state corporations must qualify with the Nevada Secretary of State to do business in the state. For information, contact:

Contact

Secretary of State
101 North Carson Street
Carson City, NV 89701-4786
(702) 687-5203
Fax: (702) 687-3471

New Hampshire

Only specialty contractors, including asbestos and lead abatement, as well as electrical and plumbing contractors, need to be licensed in New Hampshire.

Asbestos Abatement License and Certificates

If you work with asbestos in New Hampshire you must be certified and/or licensed by the New Hampshire Bureau of Health Risk Assessment. The Bureau issues an asbestos abatement contractor license, and certificates to asbestos workers and supervisors. It also certifies inspectors, management planners, and project designers who work on school buildings.

To get an application for a certificate or license, contact:

Contact

New Hampshire Division of Public Health Services
Bureau of Health Risk Assessment
Asbestos Management and Control Program
Health and Welfare Building
6 Hazen Drive
Concord, NH 03301-6527
(603) 271-4609
(800) 852-3345 (New Hampshire only)

Asbestos Abatement Contractor's License

To qualify for a contractor's license you must:

- have access to at least one state-approved asbestos disposal site
- read and understand New Hampshire Asbestos Management Rules
- provide documentation of completion of an approved training course for asbestos abatement contractors and supervisors
- list all names or other identifiers you've ever done business under
- list any outstanding enforcement actions pending against you for asbestos abatement work
- list any New Hampshire certified asbestos abatement supervisors you employ

Fee

Asbestos abatement contractor's license fee: The contractor's license will cost you $1,000 and it's good for a year.

Asbestos Abatement Supervisor Certificate

To qualify for a supervisor certificate, you need to:

- submit a signed statement saying that you have read and understand the New Hampshire Asbestos Management Rules
- pass an approved training course which includes an exam
- have at least 12 months of asbestos abatement work experience
- list any outstanding enforcement actions pending against you for asbestos abatement work

Asbestos abatement supervisor certificate fees: The supervisor's certificate will cost you $200 and it's good for a year. If you've been certified in another state, you can petition the Bureau to be certified in New Hampshire without repeating the training or work experience the Bureau requires. You still must pay the certification fee.

Asbestos Abatement Worker Certificate

To qualify for a worker certificate you need to:

- submit a signed statement saying that you have read and understand the New Hampshire Asbestos Management Rules
- pass an approved training course which includes an exam

Asbestos abatement worker certificate fees: The worker's certificate will cost you $50 and it's good for a year. If you've been certified in another state, you can petition the Bureau to be certified in New Hampshire without repeating the training or work experience the Bureau requires. You still must pay the certification fee.

Asbestos Abatement Inspector Certificate

To qualify for an inspector certificate you need to:

- submit a signed statement saying that you have read and understand the New Hampshire Asbestos Management Rules
- pass an approved training course which includes an exam
- have at least six months of asbestos abatement work experience or two months experience under the supervision of a certified asbestos inspector or management planner
- list any outstanding enforcement actions pending against you for asbestos abatement work

Asbestos abatement inspector certificate fees: The inspector's certificate, which is good for a year, will cost you $100. If you've been certified in another state, you can petition the Bureau to be certified in New Hampshire without repeating the training or work experience the Bureau requires. You still must pay the certification fee.

Asbestos Abatement Management Planner Certificate

To qualify for an management planner certificate you need to:

- submit a signed statement saying that you have read and understand the New Hampshire Asbestos Management Rules
- pass an approved training course which includes an exam
- have an associate degree or certificate of completion in one of the following:

  project planning   architecture

  management   industrial hygiene

  environmental sciences   occupational health

  engineering   related scientific field

  construction

- document at least six months of asbestos abatement work experience
- list any outstanding enforcement actions pending against you for asbestos abatement work

Asbestos abatement management planner certificate fees: The management planner's certificate, which costs $100, is good for a year. If you've been certified in another state, you can petition the Bureau to be certified in New Hampshire without repeating the training or work experience the Bureau requires. You still must pay the certification fee.

Asbestos Abatement Project Designer Certificate

To qualify for an project designer certificate you need to:

- submit a signed statement saying that you have read and understand the New Hampshire Asbestos Management Rules
- pass an approved project designer training course which includes an exam
- complete one of the following:

  1. twelve months asbestos abatement work experience and a bachelor's degree in industrial hygiene, occupation health, environmental science, biological science, or physical science

  2. twelve months asbestos abatement work experience and be a registered architect or engineer

  3. two years asbestos abatement work experience including asbestos abatement design

Asbestos abatement project designer certificate fees: The project designer's certificate will cost you $100 and it's good for a year. If you've been certified in another state, you can petition the Bureau to be certified in New Hampshire without repeating the training or work experience the Bureau requires. You still must pay the certification fee.

Lead Abatement License and Certificates

If you work in lead abatement in residential dwellings or child care centers in New Hampshire, you must be certified and/or licensed by the New Hampshire Bureau of Health Risk Assessment. The Bureau issues a license for lead abatement contractors and certificates for lead workers and supervisors. It also certifies inspectors, management planners, and project designers who work on school buildings.

To get an application for a certificate or license, contact:

Contact

Bureau of Health Risk Assessment
Childhood Lead Poisoning Prevention Program
Health and Welfare Building
6 Hazen Drive
Concord, NH 03301-4609
(603) 271-4607
(800) 852-3345 (New Hampshire only)

Lead Abatement Contractor's License

To qualify for a contractor's license you must:

- read and understand New Hampshire Lead Poisoning Prevention Act and Rules
- document completion of an approved training course for lead abatement contractors which includes an exam
- document two years work experience in asbestos, lead, environmental remediation or building trades
- list all names or other identifiers you've ever done business under
- list any outstanding enforcement actions pending against you for lead abatement work
- list all New Hampshire certified lead abatement workers/supervisors you employ

Fee

Lead abatement contractor's license fees: The contractor's license will cost you $250 and it's good for a year.

Lead Abatement Supervisor Certificate

To qualify for a supervisor certificate you need to:

- read and understand New Hampshire Lead Poisoning Prevention Act and Rules
- pass an approved training course which includes an exam
- have at least twelve months of lead abatement work experience
- have at least twelve months work experience in environmental remediation or building trades
- list any outstanding enforcement actions pending against you for lead abatement work

Lead abatement supervisor certificate fees: The supervisor's certificate will cost you $100 and it's good for a year. If you've been certified in another state, you can petition the Bureau to be certified in New Hampshire without repeating the training or work experience the Bureau requires. You still must pay the certification fee.

Lead Abatement Worker Certificate

To qualify for a worker certificate you need to:

- read and understand New Hampshire Lead Poisoning Prevention Act and Rules
- pass an approved training course which includes an exam

Lead abatement worker certificate fees: The worker's certificate will cost you $50 and it's good for a year. If you've been certified in another state, you can petition the Bureau to be certified in New Hampshire without repeating the training or work experience the Bureau requires. You still must pay the certification fee.

Lead Inspector/Technician Certificate

To qualify for a lead inspector/technician certificate you need to:

- read and understand New Hampshire Lead Poisoning Prevention Act and Rules
- pass an approved training course which includes an exam
- have conducted at least 25 full inspections over at least three months under the supervision of a lead inspector/planner/project designer
- list any outstanding enforcement actions pending against you for lead abatement work
- have a high school diploma or equivalent

Lead inspector/technician certificate fees: The inspector/technician's certificate will cost you $200 and it's good for a year. If you've been certified in another state, you can petition the Bureau to be certified in New Hampshire without repeating the training or work experience the Bureau requires. You still must pay the certification fee.

Lead Inspector/Planner/Project Designer Certificate

To qualify for a lead inspector/planner/project designer certificate you need to:

- read and understand New Hampshire Lead Poisoning Prevention Act and Rules
- pass an approved training course which includes an exam
- have conducted at least 15 full inspections and 10 clearance inspections over at least three months as a lead inspector/technician

You also need one of the following:

- certification as an industrial hygienist; architect; civil, environmental, or structural engineer; historical architect; or architectural historian
- an associate degree or certificate of completion in one of the following:

project planning	architecture
management	industrial hygiene
environmental sciences	occupational health
engineering	related scientific field
construction	

- documentation of at least six months of lead abatement work experience
- a listing of any outstanding enforcement actions pending against you for lead abatement work

Fee

Lead inspector/planner/project designer certificate fees: The management planner's certificate will cost you $100 and it's good for a year. If you've been certified in another state, you can petition the Bureau to be certified in New Hampshire without repeating the training or work experience the Bureau requires. You still must pay the certification fee.

Project Designer Certificate

To qualify for an project designer certificate you need to:

- submit a signed statement saying that you have read and understand the New Hampshire Lead Poisoning Prevention Act and Rules
- pass an approved project designer training course which includes an exam
- complete one of the following:

  1. a bachelor's degree in industrial hygiene, occupational health, environmental science, biological science, or physical science and 12 months lead abatement work experience

  2. be a registered architect or engineer and 12 months lead abatement work experience

  3. two years of lead abatement work experience including lead abatement design

Fee

Project designer certificate fees: The project designer's certificate will cost you $100 and it's good for a year. If you've been certified in another state, you can petition the Bureau to be certified in New Hampshire without repeating the training or work experience the Bureau requires. You still must pay the certification fee.

Electrician's Licenses

You need a license to do electrical work in New Hampshire. To get an application for a license, contact:

Contact

Office of the State Fire Marshal
Electrician's Licensing Board
78 Regional Drive, Bldg. 1
P.O. Box 646
Concord, NH 03302-0646
(603) 271-3748

The Board issues master and journeyman licenses and apprentice identification cards. They also require you to pass an exam to get a license.

To qualify for the master's licensing exam you must work as a New Hampshire-licensed journeyman for at least one year.

To qualify for the journeyman's licensing exam you need four years of work experience and an Associate Degree in Electricity (or Board-approved equivalent). You can use your schooling for up to one year of the work experience requirement.

To qualify for an apprentice electrician identification card you need to register as an apprentice with the Bureau of Apprenticeship and Training of the U.S. Department of Labor or a state apprenticeship agency. The Board may accept out-of-state work experience and education for any of these requirements.

The journeyman and master exams have two sections. One section has 25 questions based on the *National Electrical Code*. The other section has 50 questions based on:

- RSA 319-C

- Practical electrical installations as defined in RSA 319-C:12, III

- *American Electricians Handbook*, 1996, 13th edition, Croft/Summers, McGraw-Hill Inc., Box 543, Blacklick, OH 43004-0543

If you have a license in Massachusetts, Vermont, or Maine, you can apply for a New Hampshire license by reciprocity. You'll still need four years of work experience and you have to pay the licensing fees. You can't get reciprocity if you've ever failed a New Hampshire electrician's exam.

Fee

Electrician's license fees: It will cost you $25 to file an application for an electrical exam. The master license costs $60 and the journeyman license costs $20. Your license will expire on the last day of the month of your birth.

Plumber's Licenses

You need a license to do plumbing work in New Hampshire. To get an application for a license, contact:

Contact

State Board for Licensing and Regulating Plumbers
2 Industrial Park Drive
P.O. Box 1386
Concord, NH 03302-1386
(603) 271-3267

The Board issues master, journeyman, and apprentice licenses. They also require you to pass an exam to get a license. To qualify for the master's licensing exam you must work as a licensed journeyman for at least six months. To qualify for the journeyman's licensing exam you need to complete a Board-approved apprenticeship program which includes four years of work experience and four years of education. If you have a license in another state, you can apply for the journeyman exam without completing the apprenticeship program. If you have a license from the state of Vermont you can get a license in New Hampshire by just paying the New Hampshire license fees.

Fee

Plumber's license fees: It will cost you $25 to file an application for a plumbing license. The master license costs $125 and the journeyman license costs $60. A plumbing license is good for one year.

Department of Transportation (DOT)

To bid on New Hampshire Department of Transportation projects you have to be prequalified by the Department. To get an application for prequalification contact:

Contact

Department of Transportation
John O. Morton Building
4 Hazen Drive
P.O. Box 483
Concord, NH 03302-0483
(603) 271-3734
Fax: (603) 271-3914

Some of the details the Department will ask you about are your organization, personnel, financial condition, equipment, and experience. You'll be asked what type of work you want to be qualified for:

Road construction	Site work
Bridge construction	Building
Paving	Other

Prequalification is good for 15 months from the date of the financial statement you send the Department when you apply for prequalification.

Out-of-State Corporations

Out-of-state corporations working in New Hampshire must be licensed by the Secretary of State. For information contact:

Contact

Secretary of State
State House, Room 204
107 North Main St.
Concord, NH 03301-4989
(603) 271-3246
Fax: (603) 271-3247

New Jersey

You must register to be in the business of building new homes in New Jersey. You must also warrant each new home you build and provide warranty follow-up services. To get an application contact:

Contact

Department of Community Affairs
Bureau of Homeowner Protection
New Home Warranty Program
Trenton, NJ 08625-0805
(609) 530-8800

The application will ask you for information about:

- principal officers of your company
- any bankruptcy proceedings your company has been involved in
- any consumer fraud proceedings your company has been involved in

Fee

Registration fee: It will cost you $200 nonrefundable to register. Registration is good for two years.

Electrical Contractor's License

You need a license to do electrical work in New Jersey. To apply for an electrical contractor's license, contact:

Contact

Board of Examiners of Electrical Contractors
124 Halsey St., 6th Floor
P.O. Box 45006
Newark, NJ 07101
(201) 504-6410

The Board does require you to pass an exam to get a license. To take the exam you must be over the age of 21, have a high school diploma or equivalent, and one of the following:

- five years experience working with tools to install, alter or repair electrical wiring for light, heat or power

- completion of a four-year Board-approved apprenticeship program and one year of Board-approved work experience
- completion of 8,000 hours of experience working with tools to install, alter or repair electrical wiring for light, heat or power and 576 hours related classroom instruction
- a bachelor's degree in electrical engineering and two years of Board-approved work experience

The Board will review your application and if you're eligible they'll send you information on the exam they require. National Assessment Institute gives the exam. For information on the exam or scheduling, you can contact NAI at:

Contact

National Assessment Institute
261 Connecticut Drive
Burlington, NJ 08016
(609) 387-7944

The electrical contractor exam has three parts — one on business and law, one on the trade, and one on alarm systems. The business and law exam is open book with 50 multiple choice questions. It lasts two hours. Here's a summary of the subjects on this part and the approximate percentage of questions on each:

Subject	Percent of exam
Contract management	15
Estimating and bidding	15
Financial management	10
Licensing	10
Project management	10
Labor laws	8
Risk management	8
Safety	8
Lien laws	6
Tax laws	6
Business organization	4

The trade exam is open book with 100 multiple choice questions. It lasts four hours. You may need to use arithmetic and/or simple algebra on this exam. There may be figures, drawings, tables, or charts. Here are the subjects on the trade exam and the approximate percentage of questions on each:

Subject	Percent of exam
Grounding, bonding	11
Services, feeders, branch circuits	12
Raceways, enclosures	11
Conductors	10

Subject	Percent of exam
Motors and controls	11
Utilization and general use equipment	12
Special occupancies and equipment	6
General knowledge of electrical trade, calculations	25
Low voltage circuits, alarms and calculations	3

The alarm systems exam is open book with 30 multiple choice questions. It lasts one and one-half hours. Here are the subjects on the alarm systems exam and the approximate percentage of questions on each:

Subject	Percent of exam
National Electrical Code	60
General electrical knowledge	15
Equipment	15
Signals transmission, conductors	5
Systems design	5

Electrician's license fees: It will cost you $100 to file an application for a license. The business and law exam will cost you $35, the trade exam is $65, and the alarm systems exam is $35. All fees are nonrefundable. You'll also have to pay $5 to rent the NEC book at the exam.

The Board issues licenses every three years. Your license will cost you $50 to $150, depending on when you get it during the three year period.

Before doing any business in New Jersey you must post a bond to the state for $1,000. You must also carry $300,000 property damage and bodily injury insurance and $300,000 combined property damage and bodily injury insurance.

Recommended Reading for the Electrical Contractor's Exams

Business and Law Exam

- *NJ Electricians Reference Manual for Contractors*

- *NJ Licensing Regulations - N.J.S.A. 45:5A-1 to 45:5A-22, 45:1-14 to 45:1-27, and N.J.A.C. 13:31-1.1 to 13:31-1.16*

- *Fair Labor Standards Act*

- *INS Form I-9*

- *NJ employment regulations (Wage & Hour Law, Child Labor Act and Worker's Compensation)*

- *Circular E, Employer's Tax Guide*

Recommended Reading for the Electrical Contractor's Exams (continued)

Business and Law Exam (continued)

📖 *New Jersey Withholding Tables, NJ-WT*

📖 *Code of Federal Regulations - Title 29, Part 1926 (OSHA)*, July 1995, Superintendent of Documents, U.S. Government Printing Office, Washington, DC 20402

📖 *New Jersey Lien Laws*

📖 *The Americans with Disabilities Act, Your Responsibilities as an Employer - EEOC-BK-17*, U.S. Equal Employment Opportunity Commission

These references are reprinted in the New Jersey Electricians Business and Law Reference Manual. NAI sells this manual for $40.

Trade Exam

📖 *Alternating Current Fundamentals,* Duff and Herman, Delmar Publishers, P.O. Box 6904, Florence, KY 41022

📖 *Direct Current Fundamentals,* Loper and Tedsen, Delmar Publishers, P.O. Box 6904, Florence, KY 41022

📖 *Electric Motor Control,* Walter N. Alerich, Delmar Publishers, P.O. Box 6904, Florence, KY 41022

📖 *National Electrical Code Blueprint Reading*, K. L. Gebert, American Technical Publishers, 1155 West 175th Street, Homewood, IL 60403

📖 *Ferm's Fast Finder Index*, Olaf G. Ferm, Ferms Finder Index Company

📖 *American Electricians Handbook*, 1996, 13th edition, Croft/Summers, McGraw-Hill Inc., Box 543, Blacklick, OH 43004-0543

Alarm Systems Exam

📖 *NFPA 70 - National Electrical Code*, 1996 edition, National Fire Protection Association, 1 Batterymarch Park, P.O. Box 9101, Quincy, MA 02269

📖 *NFPA 72- National Fire Alarm Code*, National Fire Protection Association, 1 Batterymarch Park, P.O. Box 9101, Quincy, MA 02269

📖 *NEMA Training Manual on Fire Alarm Systems*, National Electrical Manufacturers Association

📖 *Fire Alarm Signaling Systems Handbook*, Bukowski and O'Laughlin, National Fire Protection Association, 1 Batterymarch Park, P.O. Box 9101, Quincy, MA 02269

**Recommended Reading for the
Electrical Contractor's Exams (continued)**

Alarm Systems Exam (continued)

📖 *Design and Application of Security/Fire Alarm Systems*, John Traister, McGraw-Hill Company

📖 *Basic Electronics Technology*, Howard W. Sams & Co.

📖 *Proprietary Burglar Alarm Systems, U.L. #1076*, Underwriters Laboratories, Inc.

You can get these books from:

Builders' Book Depot
1033 East Jefferson, Suite 500
Phoenix, AZ 85034
(800) 284-3434

Plumbing Contractor's License

You need a license to do plumbing work in New Jersey. To apply for a master plumber license, contact:

Contact

Board of Examiners of Master Plumbers
124 Halsey St., 6th Floor
P.O. Box 45008
Newark, NJ 07101
(201) 504-6420

The Board does require you to pass an exam to get a license. To take the exam you must be over the age of 21 and have one of the following:

- five years of work experience in plumbing with three or more of the years as a journeyman plumber
- a bachelor's degree in mechanical, plumbing, or sanitary engineering and one year of Board-approved work experience

The Board will review your application and if you're eligible they'll send you information on the exam they require. National Assessment Institute gives the exam. For information on the exam or scheduling, you can contact NAI at:

Contact

National Assessment Institute
261 Connecticut Drive
Burlington, NJ 08016
(609) 387-7944

The master plumber exam has three parts — business and law, trade, and practical. The business and law exam is open book with 50 multiple choice questions. It lasts two and one-half hours. Here are the subjects on this part and the percentage of questions on each:

Subject	Percent of exam
Project management	10
Contract management	20
Financial management	10
Estimating and bidding	10
Risk management	8
Business organization	4
Licensing	8
Safety	8
Labor laws	8
Tax laws	6
Lien laws	6

The trade exam is closed book, three hours long, with 100 multiple choice questions. You may need to use arithmetic and/or simple algebra on this exam. There may be figures, drawings, tables, or charts in the exam. Here's the information about the subjects on the trade exam and the percentage of questions on each:

Subject	Percent of exam
Drainage systems, sewers	22
Water supply, backflow prevention	20
Code, general knowledge	20
Installation practices, methods, materials	10
Special wastes, roof drains	10
Fixtures, trim	6
Excavation	6
Inspection, testing	6

Plumber's license fees: It will cost you $100 nonrefundable to file an application for a license. The business and law exam will cost you $35 nonrefundable, the trade exam is $65 nonrefundable, and the practical exam is $90 nonrefundable.

The Board issues licenses every two years. Your license will cost you $60 to $120, depending on when you get it during the two-year period.

If you can prove you have a valid license in a state with qualifications at least equal to New Jersey's and that state accepts New Jersey master plumbers licenses, you can apply for a license without taking the exam. You still need to pay the fee though.

Before doing any business in New Jersey, you must post a bond to the state for $3,000.

Recommended Reading for the Plumbing Contractor's Exams

Business and Law Exam:

📖 *NJ Reference Manual for Master Plumbers*

📖 *NJ Licensing Regulations - N.J.S.A. 45:5A-1 to 45:5A-22, 45:1-14 to 45:1-27, and N.J A.C. 13:31-1.1 to 13:31-1.16*

📖 *Fair Labor Standards Act*

📖 *NJ employment regulations (Wage & Hour Law, Child Labor Act and Worker's Compensation)*

📖 *Circular E, Employer's Tax Guide*

📖 *New Jersey Withholding Tables,* NJ-WT

📖 *Code of Federal Regulations - Title 29, Part 1926 (OSHA),* July 1995, Superintendent of Documents, U.S. Government Printing Office, Washington, DC 20402

📖 *New Jersey Lien Laws*

📖 *Occupational Safety and Health Act of 1972: Record keeping and Asbestos Abatement*

These references are reprinted in the New Jersey Reference Manual for Master Plumbers. NAI sells this manual for $40.

Trade Exam:

📖 *National Standard Plumbing Code,* NAPHCC, 1996 version

📖 *The Plumbers Handbook,* 8th edition, Joseph P. Almond, MacMillan Publishing Co.

📖 *Excavation & Grading Handbook,* 1991, Nicholas E. Capachi, Craftsman Book Company, 6058 Corte del Cedro, P.O. Box 6500, Carlsbad, CA 92018

You can get these books from:

Builders' Book Depot
1033 East Jefferson, Suite 500
Phoenix, AZ 85034
(800) 284-3434

Department of Transportation (DOT)

To bid on New Jersey Department of Transportation projects you have to be prequalified by the Department. To get an application for prequalification contact:

Contact

New Jersey Department of Transportation
1035 Parkway Avenue
CN 600
Trenton, NJ 08625-0600
(609) 530-2103
Fax: (609) 530-2238

Some of the details the Department will ask you about are your organization, personnel, financial condition, equipment, and experience. You'll be asked what type of work you want to be qualified for. Here are the types of work the Department uses:

Grading	Permanent signs
Paving	Underground utilities
Grading and paving	Waterproofing
Bridges	Maintenance, protection of traffic
Heavy highway	Milling
Landscapes	Blasting
Electrical	Removal of petroleum products, debris, hazardous material
General concrete	
Miscellaneous	Health and safety plan
Demolition	Sampling and analysis
Dewatering	Design services
Dredging	Modified design build
Fencing	Computerized arterial traffic control systems
Rest, service buildings	
Guardrails	Intelligent transportation systems
Impact attenuator installation	System integration
Standard pavement markings	Rumble strips
Long life pavement markings	Rehabilitation of movable bridge houses
Pneumatic mortar or gunite	Machine sweeping
Sand drains, sand fill	Corrective action work

If your company isn't in New Jersey, you must specify a New Jersey resident who'll accept service of any legal documents from the Department. Prequalification is good for 18 months from the date of the financial statement you send with your application.

Out-of-State Corporations

Out-of-state corporations working in New Jersey must be licensed by the Secretary of State. For information, contact:

Contact

Secretary of State
Division of Commercial Recordings
820 Bear Tavern Road CN 308
Trenton, NJ 08625
(609) 530-6400
Fax: (609) 530-6433

New Mexico

To do construction work in New Mexico you must be licensed. The state agency that issues licenses is:

Construction Industries Division
State of New Mexico Regulation and Licensing Department
725 St. Michael's Drive
P.O. Box 25101
Santa Fe, NM 87504
(505) 827-7030
Fax: (505) 827-7045

Construction Contractor's Licenses

If you request an application for a license, the Division will send you a brochure and an information sheet on applying for a contractor's license. This information gives you the Division licensing procedure and general work experience requirements. There are also some New Mexico residence requirements. Corporations must be qualified to do business in New Mexico or be incorporated in the state. Any out-of-state corporation must have a registered agent and office in New Mexico for at least 90 days before it can get a license. An individual or partnership must have a residence or street address in New Mexico for at least 90 days before applying for a license. You also must have a current New Mexico Taxation and Revenue Department tax identification number and, if appropriate, workers' compensation insurance that's been validated by the New Mexico Workers' Compensation Administration.

Before you can submit an application, you have to get an item called Kit #3 from National Assessment Institute (NAI). NAI is the private firm that handles license testing for the Division. It will cost you $13 plus tax for the kit. You can reach NAI at:

National Assessment Institute
1221 St. Francis Drive Suite B
Santa Fe, NM 87505
(505) 982-8197
Fax: (505) 986-1299

The kit contains the New Mexico Construction Industries Licensing Act, New Mexico Construction Industries Division Rules and Regulations, and an NAI information brochure on the business and law and trade exams you have to take to get a contractor's license.

You need to go through this material and pick what types of work you want to be licensed in. The trades require two or four years of work experience. Here's a list of the licenses the Division uses. The trades for which you need four years of experience are marked with an asterisk:

General construction*
- asphalt, bitumen, concrete construction
- streets, road, highways, tunnels, parking lots, alleys, seal coat, surfacing
- striping
- maintenance, repairs
- highway signs, guardrails
- curbs, gutters, culverts

Building construction
- residential
- general building*

Building specialty
- acoustical/insulation, urethane foam
- awnings, canopies
- ceramic tile, marble, terrazzo
- concrete, cement, walkways, driveways
- demolition
- door installation
- drywall
- earthmoving, excavating, ditching
- elevators, escalators, conveyers
- fencing
- fixtures, cabinets and millwork
- floor covering, seamless and wood floors, finishing
- framing
- glazing, weatherstripping, storm doors, window installation
- caissons, piers, pile driving
- masonry
- ornamental iron, welding
- painting, decorating
- remodeling

- trenching and backhoe owner/operator
- roofing
- sandblasting
- sign construction (nonelectrical)
- structural steel erection
- swimming pools (nonmechanical/electrical)
- vaults, depositories
- gunite
- construction and technical specialties
- plastering, stucco, lathing
- siding
- sheet metal
- concrete coring, drilling, slab sawing

Fixed works*

- airports
- bridges
- canals, reservoirs or irrigation systems
- drainage or flood control systems
- recreation areas
- railroad, tunnel construction
- tanks, towers
- transmission lines, tanks, substations
- utility lines

Electrical

- residential wiring
- electrical distribution systems (including transmission lines)*
- electrical (including residential wiring and electrical specialties)*

Electrical specialty

- electrical signs, outline lighting
- cathodic protection
- sound, intercommunication, electrical alarm systems
- electrical traffic control systems
- microwave communications systems
- cable tv
- telephone communication systems
- telephone interconnect systems

Mechanical*

- plumbing
- natural gas fitting
- air conditioning, ventilation
- heating, cooling, process piping

Mechanical specialty

- residential plumbing
- residential natural gas fitting
- cesspools, septic tanks, sewers
- appliance installation, service
- evaporative coolers
- lawn sprinklers
- swimming pool piping
- water conditioners
- refrigeration
- boiler installation, repair, service
- controls (mechanical systems)
- fire protection sprinkler systems
- pneumatic tube systems
- dry chemical fire protection

Journeyman classifications

- electrical
- electrical distribution systems, including transmission lines
- electrical specialties
- electrical signs, outline lighting
- cathodic protection
- sound, intercommunication, electrical alarm systems
- electric traffic control systems
- microwave communication systems
- cable TV
- telephone communications systems
- telephone interconnect systems

Mechanical

- plumber
- natural gas fitter
- boiler operator
- sheet metal

- pipefitter
- welder
- specialty

The business and law exam is open book, based on the New Mexico Contractors' Reference Manual which NAI sells for $45 plus tax and shipping. The exam is multiple choice format and lasts two hours. Here's the information:

Subject	Percent of exam
Licensing	20
Federal and state tax laws	15
Labor laws	11
Estimating and bidding	10
Contract management	8
Project management	7
Risk management	7
Financial management	7
Business organization	5
Lien laws	5
Safety	5

You can substitute a Division-approved business and law course for this exam. It will cost you $31.88 to take it.

The trade exams are also given by NAI. They're based on surveys of licensed contractors in New Mexico on:

- accepted industry practices
- how to read and interpret construction codes, regulations and building codes
- trade materials, tools, equipment, methods

The trade exams are multiple choice, but the following trades also have a written essay:

Acoustical/insulation, urethane foam

Awnings, canopies

Demolition

Door installation

Fixtures, cabinets, millwork

Floor covering, seamless and wood floors, finishing

Cassions, piers, pile driving

Ornamental iron, welding

Trenching and backhoe owner/operator

Sandblasting

Sign construction (nonelectrical)

Swimming pools (nonmechanical/electrical)

Vaults, depositories

Gunite

Various specialties

Concrete coring, drilling, slab sawing

NAI is in the process of replacing the essay parts of these trade exams with multiple choice questions. You can get a content outline for most trade exams from NAI.

Construction contractor's exam fees: A trade exam will cost you $31.88 except for the following trades whose exams have multiple parts. Here's a summary of the trade exams, number of parts in each, and their cost:

Trade	Number of parts in exam	Cost of exam
Asphalt, bitumen, concrete construction	5	$159.38
Fixed works	9	$286.88
Residential building	2	$63.75
Building construction	4	$127.50
Mechanical	4	$127.50
Electrical	3	$95.64

Construction contractor's license fees: It will cost you $30 nonrefundable to apply for your initial contractor's license. The license itself will cost $50 for each trade except for the trades which have multipart exams. Licenses for these trades are $100 each. Journeyman exams and licenses cost $25. Licenses are good for one year.

New Mexico doesn't accept any license you have in another state.

Out-of-State Corporations

Out-of-state corporations should register with the New Mexico Secretary of State to do business in the state. Contact them at:

New Mexico Secretary of the State
Corporation Commission
P.O. Drawer 1269
Santa Fe, NM 87504
(505) 827-4504
Fax: (505) 827-4387

New York

Except for asbestos abatement work, all construction work in New York is regulated at the local level.

Asbestos Abatement Contractor's License

To do asbestos abatement work in New York you must get a license. To get an application, contact:

Contact

Division of Safety and Health
License and Certificate Unit
Mail Stop 7G
P.O. Box 687
New York, NY 10014-0687
(212) 352-6106
Fax: (212) 352-6186

Fee

Asbestos abatement contractor's license fees: To get a contractor's license you must certify that all your employees are certified to do asbestos work. You must also have workers' compensation and disability insurance. It will cost you $300 to get an asbestos abatement contractor's license. The license is good for one year. New York doesn't have any reciprocity agreements with other states for this license.

The Division also issues individual certificates. Here's a list of them and the cost of each:

Certificate type	Fee
Handler	$30
Restricted handler	$30
Operations and maintenance	$30
Air sampling technician	$50
Inspector	$100

Management planner	$150
Supervisor	$50
Project designer	$150
Project monitor	$150

You must have passed an approved asbestos training course for any certificate no more than one year before you apply for the certificate. An individual certificate is good for one year but it expires on the last day of your birth month.

Department of Transportation (DOT)

You don't have to be prequalified to bid on New York Department of Transportation projects. However you must fill out a questionnaire from the Contracts Management Bureau if you're a low bidder on a bid contract. Some of the details the Department will ask you about are your organization, personnel, financial condition, equipment, and experience. If you submit this questionnaire to the Department and the information doesn't change, it's good for one year. The Bureau address is:

Contact

Contracts Management Bureau
Department of Transportation
Washington Ave., State Campus
Albany, NY 12232
(518) 457-3583
Fax: (518) 457-8475

Proposed projects are listed in the Notice of Highway Lettings. You can subscribe to this for $32 a year. For subscription information, contact the Bureau.

The Department has requirements for its construction projects that you need to know about before bidding on a project. You can get information about this in their publication, *Standard Specifications*, which they sell for $25.

Out-of-State Corporations

Corporations doing business in New York must register with the New York Secretary of State to do business in the state. For information, contact:

Division of Corporations
Department of State
162 Washington Avenue
Albany, NY 12231
(518) 473-2278
Fax: (518) 474-5173

North Carolina

To work as a general contractor on projects costing more than $30,000 in North Carolina, you must get a license from the North Carolina Licensing Board for General Contractors. To get an application, contact:

Contact

North Carolina Licensing Board for General Contractors
P. O. Box 17187
Raleigh, NC 27619
(919) 571-4183
Fax: (919) 571-4703

If your company is a partnership or corporation you must register with the North Carolina Secretary of State and include the registration with your application to the Board. You can contact the Secretary of State at:

Contact

North Carolina Secretary of State
300 North Salisbury St.
Raleigh, NC 27611
(919) 733-4201
Fax: (919) 733-1837

You must furnish proof of financial responsibility on your application. The Board issues a limited, intermediate, or unlimited license according to a company's working capital. Working capital is current assets minus current liabilities. Here's the working capital the Board requires for each license class:

License class	Minimum required working capital	Cost per project
Limited	$12,500	less than $250,000
$50,000	$50,000	less than $500,000
Unlimited	$100,000	no limit

You must also include five letters of reference that are less than six months old with your application.

General Contractor's License

The Board issues several types of contractor licenses. You must pass an exam to get any type of license. Here are the types of contractor licenses the Board uses:

- Building
- Residential
- Highway
- Public utilities
- Specialty

Specialty contractor licenses include:

Water and sewer lines	Boring and tunneling
Water purification and sewage disposal	Concrete construction
Grading and excavating	Interior construction
Asbestos	Marine construction
Insulation	Masonry construction
Roofing	Railroad construction
Electrical (ahead of point of delivery)	Metal erection
Communications	Swimming pools
Fuel distribution	

You can also get an Unclassified license by passing the Building, Highway, and Public Utilities exams. Here's a basic outline of the content of each exam.

Business management, operations fundamentals:

- Accounting terminology
- Department of Labor wage, hour requirements
- Workers' compensation
- Employment Security Commission requirements

Laws and regulations:

- North Carolina general contracting laws, regulations
- OSHA regulations, safety requirements
- Liens, construction laws
- Warranties

Code:

- Familiarity with applicable code

Technology:

- Basic surveying
- Drywall, lath, plaster
- Foundation, footings, brick, masonry, site work

- Description of materials and their use
- Steel construction
- Finish work
- Insulation, particleboard, other structural members installation
- Rough carpentry

The Building, Residential, and Roofing exams have an additional section on blueprint reading. This section covers:

- Understanding specifications
- Take-off materials
- Calculating materials
- Figuring quantity and extending material, labor costs
- Heating, plumbing, and electrical basic knowledge
- Site, grading, and landscape plan and detail

If you're taking the Swimming Pool exam, you may have to complete a take-off on a pool.

General contractor's license fees: It costs $50 nonrefundable for each exam you apply for. It costs $100 to apply for an unlimited license. The application fee for an intermediate license is $75 and the limited license application fee is $50. The application fee includes the first year's license fee. All licenses expire on December 31 each year.

If you have a license in another state whose licensing requirements are essentially equal to the Board's, you can apply for a similar license in North Carolina. You won't have to take an exam but you will have to pay whatever fee the Board requires for the license.

Recommended Reading for General Contractor's Exams

All Contractor's Exams:

- *North Carolina Occupational Safety and Health Standards for the Construction Industry*, North Carolina Department of Labor, 319 Chapanoke Road, Raleigh, NC 27603

- *North Carolina Workers' Compensation Act*, North Carolina Industrial Commission, 430 North Salisbury St., Raleigh, NC 27611

- *Employment Security Law of North Carolina*, Employment Security Commission of North Carolina, 700 Wade Ave., Raleigh, NC 27605

Recommended Reading for General Contractor's Exams (continued)

Building:

📖 *North Carolina State Building Code Volume 1, 1-A, and 1-C*, North Carolina Department of Insurance Engineering Division, P. O. Box 26387, Raleigh, NC 27611

📖 *Gypsum Construction Handbook*, 1992, United States Gypsum Company, 125 South Franklin St., P. O. Box 806278, Chicago, IL 60680-4124

📖 *Ramsey-Sleeper Architectural Graphic Standards Manual*, 9th edition, John Wiley & Sons, Inc., 1 Wiley Drive, Somerset, NJ 08875

Residential:

📖 *North Carolina State Building Code Volume 7*, North Carolina Department of Insurance Engineering Division, P. O. Box 26387, Raleigh, NC 27611

📖 *Gypsum Construction Handbook*, 1992, United States Gypsum Company, 125 South Franklin St., P. O. Box 806278, Chicago, IL 60680-4124

📖 *Ramsey-Sleeper Architectural Graphic Standards Manual*, 9th edition, John Wiley & Sons, Inc., 1 Wiley Drive, Somerset, NJ 08875

Highway:

📖 *Standard Specifications for Roads and Structures*, North Carolina Department of Transportation, 1020 Birch Ridge Dr., Raleigh, NC 27610

Water and sewer lines:

📖 *North Carolina Administrative Code*, Title 15A, Department of Environmental Health & Natural Resources, Subchapter 18C, Water Supplies Rules Governing Public Water Supply Systems, North Carolina Department of Environmental Health and Natural Resources, P. O. Box 29536, Raleigh, NC 27626

Communications:

📖 *NFPA 70 - National Electrical Code*, 1996 edition, National Fire Protection Association, 1 Batterymarch Park, P. O. Box 9101, Quincy, MA 02269

📖 *Cableman's and Lineman's Handbook*, Kurtz and Shoemaker, McGraw-Hill Company, Blue Ridge Summit, PA 17294

Recommended Reading for General Contractor's Exams (continued)

Asbestos:

For general asbestos information, contact:

Health Hazard Control Branch
P. O. Box 27687
Raleigh, NC 27611-7687

Concrete construction:

📖 *Ramsey-Sleeper Architectural Graphic Standards Manual,* 9th edition, John Wiley & Sons, Inc., 1 Wiley Drive, Somerset, NJ 08875

Grading and excavating:

📖 *Standard Specifications for Roads and Structures,* North Carolina Department of Transportation, 1020 Birch Ridge Dr., Raleigh, NC 27610

Electrical (ahead of point of delivery):

📖 *NFPA 70 - National Electrical Code,* 1996 edition, National Fire Protection Association, 1 Batterymarch Park, P. O. Box 9101, Quincy, MA 02269

📖 *Cableman's and Lineman's Handbook,* Kurtz and Shoemaker, McGraw-Hill Company, Blue Ridge Summit, PA 17294

Insulation:

📖 *North Carolina State Building Code Volume 1,* North Carolina Department of Insurance Engineering Division, P. O. Box 26387, Raleigh, NC 27611

📖 *Gypsum Construction Handbook,* 1992, United States Gypsum Company, 125 South Franklin St., P. O. Box 806278, Chicago, IL 60680-4124

Fuel distribution:

📖 *ANSI B31.4 - Pressure Piping Liquid Petroleum Transportation Piping Systems,* 1974, American Society of Mechanical Engineers, United Engineering Center, 345 East 47th St., New York, NY 10017

📖 *Recommended Practices for the Installation of Underground Liquid Storage Systems,* Petroleum Equipment Institute, P. O. Box 2380, Tulsa, OK 74101

📖 *NFPA 30, 30A, National Fire Protection Association,* 1 Batterymarch Park, P. O. Box 9101, Quincy, MA 02269

📖 *ANSI B31.8 - Pressure Piping Gas Transmission and Distribution Piping Systems,* 1975, American Society of Mechanical Engineers, United Engineering Center, 345 East 47th St., New York, NY 10017

Recommended Reading for General Contractor's Exams (continued)

Roofing:

    📖 *NRCA Roofing and Waterproofing Manual*, 1996, 4th edition, National Roofing Contractor's Association, 10255 W. Higgins, Rd., Suite #600, Rosemont, IL 60018-5607

Interior construction:

    📖 *North Carolina State Building Code Volume 1*, North Carolina Department of Insurance Engineering Division, P. O. Box 26387, Raleigh, NC 27611

    📖 *Gypsum Construction Handbook*, 1992, United States Gypsum Company, 125 South Franklin St., P. O. Box 806278, Chicago, IL 60680-4124

Boring and tunneling:

    📖 *Earth Tunneling with Steel Supports, Proctor and White, Commercial Shearing, Inc.*, 1775 Logan Ave., Youngstown, OH 44501

Masonry construction:

    📖 *Ramsey-Sleeper Architectural Graphic Standards Manual*, 9th edition, John Wiley & Sons, Inc., 1 Wiley Drive, Somerset, NJ 08875

Swimming pools:

    📖 *Public Rules Government Swimming Pools 15A NCAC 18A.2500*, North Carolina Department of Environmental Health and Natural Resources, P. O. Box 29536, Raleigh, NC 27626

Fire Protection Sprinkler Contractor's License

To be a Fire Protection Sprinkler contractor you must get a license from the State Board of Examiners of Plumbing, Heating, and Fire Sprinkler Contractors. To get an application, you can contact them at:

State Board of Examiners of Plumbing, Heating, and Fire Sprinkler Contractors
3801 Wake Forest Road, Suite 201
Raleigh, NC 27609
(919) 875-3612 / Fax: (919) 875-3616

To be licensed as a contractor you must have a current certificate from the National Institute for Certificate in Engineering Technology (NICET) Level III.

Fire protection sprinkler contractor's license fees: It will cost you $75 nonrefundable to file an application. The Certificate will cost you $275 and it's good until the last day of December each year following the year you get it.

Electrical Contracting Licenses

To do electrical contracting work in North Carolina you must get a license from the State Board of Examiners of Electrical Contractors. To get an application, contact:

Contact

State Board of Examiners of Electrical Contractors
1200 Front St., Suite 105
P. O. Box 18727
Raleigh, NC 27619
(919) 733-9042 / Fax: (919) 733-6105

The Board issues Limited, Intermediate, and Unlimited licenses and these special restricted licenses:

Single-family detached residences

Low voltage

Plumbing and heating

Electric sign

Elevator

Groundwater pump

Swimming pool bonding

All licenses expire on June 30.

If you have a license in another state whose licensing requirements are essentially equal to the Board's you can apply for a similar license in North Carolina. You won't have to take an exam but you will have to pay whatever fee the Board requires for the license.

Recommended Reading for the Electrical Contractor's Exams

📖 *NFPA 70 - National Electrical Code*, 1996 edition, National Fire Protection Association, 1 Batterymarch Park, P. O. Box 9101, Quincy, MA 02269

📖 *NFPA 72- National Fire Alarm Code*, current edition, National Fire Protection Association, 1 Batterymarch Park, P. O. Box 9101, Quincy, MA 02269

📖 *Construction Management Guide - North Carolina Edition*, State Board of Examiners of Electrical Contractors, 1200 Front St., Suite 105, Raleigh, NC 27619

📖 *Laws on electrical contracting in North Carolina Administrative Code*, Chapter 18B, North Carolina General Statues, Chapter 87, Article 4, and North Carolina State Board of Examiners of Electrical Contractors Rules, Title 21.

Department of Transportation (DOT)

You have to be prequalified to bid on North Carolina Department of Transportation projects. To get an application, contact:

Contact

North Carolina Department of Transportation
Contractual Services Unit
P. O. Box 25201
Raleigh, NC 27611
(919) 733-7174
Fax: (919) 733-4141

Some of the details the Department will ask you about are your organization, personnel, financial condition, equipment, and experience. If you want to bid on projects over $50,000, you must prove your company can obtain a payment and performance bond. Also to bid on projects over $30,000 you need a license from the North Carolina Licensing Board for Contractors.

Out-of-State Corporations

Out-of-state corporations must get a Certificate of Authority from the North Carolina Secretary of State to do business in North Carolina. To apply for this certificate, contact:

Contact

North Carolina Secretary of State
300 North Salisbury St.
Raleigh, NC 27611
(919) 733-4201
Fax: (919) 733-1837

North Dakota

You must have a license in North Dakota to work on any job costing $2,000 or more. To get an application, contact:

Contact

Secretary of State
State of North Dakota
600 East Boulevard Avenue
Bismarck, ND 58505-0500
(701) 328-3665
Fax: (701) 328-1690

The state issues four classes of contractor's licenses. Here's a summary of the classes, the single contract cost limitation for each class, and its license fee:

Class	Single contract cost limitation	Fee
A	unlimited	$300
B	$250,000	$200
C	$120,000	$150
D	$50,000	$50

You don't have to take an exam to get a contractor's license. But you do have to prove you don't owe any taxes and that you have workers' compensation and liability insurance.

On the application you'll also be asked to specify your main type of work. However, you're not limited to that type of work. Here are the types of work you can choose from:

Single family
Building materials, lumber
Roofing and siding
Asbestos abatement
Paint, wall coverage
Road and related construction
Remodeler
Carpentry

Well drilling
Land developer
Flooring
Piping
Architect, planner designer, engineer
Landscaping
Excavating
Product manufacturers

Masonry   Plumbing, HVAC
Commercial developer   Other
Electrical

A license is good for one year.

Electrician's Licenses

To do electrical work in North Dakota, you need to be licensed as a master, journeyman, or class B electrician. To get an application, contact:

North Dakota State Electrical Board
P.O. Box 857
Bismarck, ND 58502
(701) 328-9522
Fax: (701) 328-9524

Contact

You must pass an exam to get a license. To qualify for the master exam you need one year of work experience as a licensed journeyman electrician. To qualify for the journeyman exam you need four years of work experience. You can use two years of Board-approved electrical education for one year of the work experience requirement. To qualify for the Class B electrician's license you need 18 months of work experience in farmstead or residential wiring. You can use two years of Board-approved electrical education for six months of the work experience requirement.

Fee

Electrician's license fees: The master exam and the license cost $50 each. The journeyman exam and license are $25 each. The Class B electrician exam and the license cost $40 each. All licenses expire March 31 each year.

The Board has some arrangements for reciprocal master or journeyman licenses. If you have a valid journeyman license in Idaho, Oregon or Washington, you can get a reciprocal license in North Dakota. If you've had a valid license in Minnesota or South Dakota for one year, you can also get a reciprocal license in North Dakota.

If you have a valid electrical license in another state which you got through a licensing procedure at least equal to North Dakota's, you can get a similar North Dakota license without taking an exam. You still have to pay the appropriate application and licensing fees, however.

Plumber's Licenses

To do plumbing work in North Dakota, you need to be licensed as a journeyman or master plumber. To get an application, contact:

Contact

North Dakota State Plumbing Board
204 West Thayer Avenue
Bismarck, ND 58501
(701) 328-9979
Fax: (701) 328-9979
E-mail: ndplumb@btigate.com

You must pass an exam to get a license. The exam is based on the North Dakota State Plumbing Code. To qualify for the master exam you need two years of work experience as a licensed journeyman plumber in North Dakota or any other state that has a state licensing law. If you have a valid master license in another state, you can use it to apply for a master exam.

Fee

Plumber's licenses fees: The exam will cost you $35. The license costs $165 and it's good for one calendar year.

To qualify for the journeyman exam you need four years of work experience under the supervision of a licensed master plumber. If you have a valid journeyman license in another state, you can use it to qualify for the exam.

You can also qualify for the journeyman exam by passing a screening exam first if you have five years of plumbing work experience in a state which doesn't license plumbers.

The exam will ask you questions about basic plumbing principles and the state plumbing code. You'll be asked to draw stacks, wastes, vents, and minimum pipe sizes on drawings which already have plumbing fixtures on them. It will cost you $25 to take the journeyman exam. The license costs $75 and it's good for one calendar year.

The Board has reciprocity agreements with South Dakota, Minnesota, and Montana.

Department of Transportation (DOT)

To bid on North Dakota Department of Transportation projects you have to be prequalified by the Department. To get an application for prequalification, contact:

Contact

North Dakota Department of Transportation
Construction Services Division
608 East Boulevard Avenue
Bismarck, ND 58505-0700
(701) 328-2563
Fax: (701) 328-4928

Some of the details the Department will ask you about are your organization, personnel, financial condition, equipment, and experience. You'll be asked what type of work you want to be qualified for. Here are the major types of work the Department uses:

General (all types of work)	Bituminous cold mix
Grading	Bituminous hot mix
Culverts, small structures	Bituminous seal coat
Gravel	Portland cement concrete pavement
Treated base	Bridge

The Department also uses these incidental construction types of work:

Riprap	Traffic signs
Water and sewer	Electrical lighting, traffic signals
Curb and gutter, sidewalk, driveways	Rest area, miscellaneous building
Painting rail	Jacking and baring pipe
Erosion control	Grind concrete pavement, saw and seal joints
Landscaping	Milling bituminous pavement
Fencing	Other
Pavement marking	

You don't need a North Dakota contractor's license to bid on a project, but if you win the contract you must get the appropriate North Dakota contractor license within ten days. Prequalification is good for one year.

Out-of-State Corporations

Out-of-state corporations must get a Certificate of Authority to do business in North Dakota from the North Dakota Secretary of State. To apply for this certificate, contact:

Contact

North Dakota Secretary of State
Main Capitol Building, first floor
600 East Boulevard Avenue
Bismarck, ND 58505
(701) 328-2939
Fax: (701) 328-2992

Ohio

The state of Ohio doesn't license contractors. That's done by the municipality where work is done. However the Ohio Construction Industry Examining Board issues Qualification Certificates for plumbing, electrical, HVAC, hydronics, and refrigeration contractors. To apply for a certificate, you need at least two years of experience in the field you're applying for. You also have to be a U. S. citizen. To get an application for a certificate, contact:

Contact

Ohio Department of Commerce
Ohio Construction Industry Examining Board
6606 Tussing Road
P.O. Box 4009
Reynoldsburg, OH 43068-9009
(614) 644-3493

The Board will review your application and if you're eligible they'll send you information on the exam they require. The exam has two parts — one on business and law and the other on HVAC, electrical, plumbing, refrigeration, or hydronics. Everyone must take the business and law part. It's a one-hour, open book exam with 30 multiple choice questions. Here's a breakdown of the number of questions on each subject:

Subject	Number of questions
Contract management	6 - 8
Project management	3 - 5
Financial management	2 - 4
Tax laws	3 - 5
Insurance and bonding	1 - 3
Business organization	0 - 2
Personnel regulations	4 - 6
Lien laws	0 - 2
Licensing	2 - 4

Here's a summary of the trade exams, including their duration and number of questions:

Trade	Duration (hours)	Number of questions
Electrical	6	150
HVAC	6	110
Hydronics	4	50
Plumbing	6	170
Refrigeration	4	50

All the trade exams are open book except for one hour of the electrical exam. They last from four to six hours depending on the trade. You're only allowed to take one trade exam per session. Here's a breakdown of subject areas and questions for each trade exam. The questions on the exam come from the references, and you're allowed to use any of them for the open book part of the exam.

Electrical Contractor's Exam

Part I (closed book, one hour, 50 questions) — 21% of total grade:

Subject	Number of questions
General theory	24 - 26
NEC chapter 1	6 - 8
NEC chapter 2	1 - 3
NEC chapter 3	0 - 2
NEC chapter 90	0 - 2
NEC chapter 4	6 - 8
NEC chapter 5	2 - 4
NEC chapter 6	2 - 4

Part II (open book, two hours, 70 questions) — 29% of total grade:

Subject	Number of questions
General theory	8 - 10
NEC chapter 1	1 - 3
NEC chapter 2	9 - 11
NEC chapter 3	12 - 14
NEC chapter 4	16 - 18
NEC chapter 5	4 - 6
NEC chapter 6	7 - 9
NEC chapter 7	0 - 2
NEC chapter 8	1 - 3
NEC chapter 9	2 - 4

Part III (open book, three hours, 30 questions) — 50% of total grade:

Subject	Number of questions
Service	4 - 6
Voltage drop	2 - 4
3Ø 208V delta and wye loads	2 - 4
Conduit fill	3 - 5
Conductor ampacity	2 - 4
Motors (branch circuit)	0 - 2
Motors (feeders)	0 - 2
Motors (protection)	0 - 2
Motors (control)	2 - 4
Appliance loads	1 - 3
Transformers, 3Ø transformer calculation	1 - 3
Grounding conductor and equipment ground	0 - 2
Power factor, power, and volt-amps	0 - 2

HVAC Contractor's Exam

Part I (open book, three hours, 50 questions) — 50% of total grade:

Subject	Number of questions
Code compliance	10 - 15
Fuel gas systems	5 - 10
Ducts	8 - 12
HVAC load calculation	8 - 12
Safety	2 - 4
Ducts (safety and fire prevention)	2 - 4

Part II (open book, three hours, 60 questions) — 50% of total grade:

Subject	Number of questions
HVAC general	10 - 15
HVAC maintenance	8 - 12
HVAC controls	8 - 12
Safety — HVAC	4 - 6
Piping	8 - 12
Insulation	4 - 8

Hydronics Contractor's Exam

Open book, four hours, 50 questions:

Subject	Number of questions
Heat loss	8 - 12
Piping	7 - 15
Equipment sizing	4 - 6
Safety	2 - 4
Controls	2 - 4
Service/maintenance	4 - 6
Code compliance	7 - 8

Plumbing Contractor's Exam

Part I (open book, two hours, 100 questions) — 40% of total grade:

Subject	Number of questions
Administration	8 - 12
Definitions	3 - 6
General regulations	10 - 14
Materials	3 - 6
Joints and connections	3 - 6
Drainage systems	8 - 12
Indirect wastes	3 - 6
Storm drains	2 - 4
Vents and venting	8 - 12
Traps	3 - 6
Interceptors	3 - 6
Fixtures	8 - 12
Hangers and supports	3 - 6
Water supply and distribution	8 - 12

Part II (open book, one hours, 30 questions) — 20% of total grade:

Subject	Number of questions
General knowledge	6 - 10
Drainage calculations	1 - 3
Vent calculations	2 - 4
Developed length	5 - 10
Fitting identification	8 - 12

Part III (open book, three hours, 40 questions) — 40% of total grade:

Subject	Number of questions
Isometric analysis	40

Refrigeration Contractor's Exam

Open book, four hours, 50 questions:

Subject	Number of questions
Code compliance	5 - 10
Refrigeration general	7 - 12
Refrigeration maintenance	5 - 10
Refrigeration controls	5 - 10
Safety — refrigeration	5 - 10
Piping — general	2 - 5
Piping — dimension and design	5 - 10
Safety	0 - 5

For more information on the exams or practice exams, contact:

Contact

Block & Associates
4442 Professional Parkway
Groveport, OH 43125
(614) 836-7881

Fee

Contractor's qualification certification exam fees: It'll cost you $25 nonrefundable to file an application to take an exam. The exams cost $50 each. Once you pass both parts of the exam, you send a copy of your scores with $25 to the Board and they'll issue your certificate. It's good for at least 12 months.

Recommended Reading for the Contractor's Exams

Electrical Contractor's Exam

📖 *NFPA 70 - National Electrical Code,* 1996 edition, National Fire Protection Association, 1 Batterymarch Park, P.O. Box 9101, Quincy, MA 02269

📖 *American Electricians' Handbook*, 11th or 12th edition, Croft, Watt, and Summers, McGraw-Hill Company

HVAC Contractor's Exam

📖 *Ohio Basic Mechanical Code,* 1995, BOCA International, 1245 S. Sunbury Road, Suite 100, Westerville, OH 43081-9308

Recommended Reading for the Contractor's Exams (continued)

HVAC Contractor's Exam (continued)

📖 *Code of Federal Regulations - Title 29*, Part 1926 (OSHA), July 1995, Superintendent of Documents, U.S. Government Print Office, Washington, DC 20402

📖 *NFPA 54 - National Fuel Gas Code*, 1992, National Fire Protection Association, 1 Batterymarch Park, P.O. Box 9101, Quincy, MA 02269

📖 *NFPA 90A - Installation of Air Conditioning and Ventilating Systems*, 1993, NFPA, 1 Batterymarch Park, Box 9101, Quincy, MA 02269-9101

📖 *ANSI/ASHRAE 15-94 - Safety Code for Mechanical Refrigeration,* 1994 ANSI, 11 West 42nd St., New York, NY 10036

📖 *Refrigeration and Air Conditioning*, 1995, 3rd edition, ARI, Prentice Hall, P.O. Box 11071, Des Moines, IA 50336-1071

📖 *Electric Controls for Refrigeration and Air Conditioning*, 1988, 2nd edition, B.C. Langley

📖 *Manual J - Load Calculation for Residential Winter and Summer Air Conditioning*, 1986, 7th edition, Air Conditioning Contractors of America, 1712 New Hampshire Ave., NW, Washington, DC 20009

📖 *Manual N - Load Calculation for Commercial Summer and Winter Air Conditioning*, 1988, 4th edition, Air Conditioning Contractors of America, 1712 New Hampshire Ave., NW, Washington, DC 20009

📖 *Manual D - Residential Duct Systems,* 1995, Air Conditioning Contractors of America, 1712 New Hampshire Ave., NW, Washington, DC 20009

📖 *System Design Manual - Part III, Piping Design*, Carrier, 1960, Carrier Air Conditioning Company, Carrier Parkway, P.O. Box 4808, Syracuse, NY 13221

📖 *Pipefitters Handbook*, 1967, 3rd edition, Forest Lindsey, Industrial Press, Inc., 200 Madison Avenue, New York, NY 10016

📖 *Trane Air Conditioning Manual*, 1965, Trane Company, 8929 Western Way, Suite #1, Jacksonville, FL 32256

📖 *Trane Ductulator*, 1976, Trane Company, 8929 Western Way, Suite #1, Jacksonville, FL 32256

📖 *HVAC Duct Construction Standards - Metal and Flexible*, 1995, Sheet Metal and Air Conditioning Contractor's National Assoc., Box 221230, Chantilly, VA 22022-1230

📖 *Fibrous Glass Duct Construction Standards*, 1993, 2nd edition, North American Insulation Manufacturers Assoc., 44 Canal Center Plaza, Suite #310, Alexandria, VA 22314

Recommended Reading for the Contractor's Exams (continued)

Hydronics Contractor's Exam

📖 *Ohio Basic Mechanical Code*, 1995, BOCA International, 1245 S. Sunbury Road, Suite 100, Westerville, OH 43081-9308

📖 *Ohio Administrative Code*, Chapter 4101:4 - Boiler and Unfired Pressure Vessel Rules, 1992, BOCA International, 1245 S. Sunbury Road, Suite 100, Westerville, OH 43081-9308

📖 *NFPA 8501 - Single Burner Boiler Operation*, 1992, National Fire Protection Association, 1 Batterymarch Park, P.O. Box 9101, Quincy, MA 02269

📖 *Code of Federal Regulations - Title 29,* Part 1926 (OSHA), July 1995, Superintendent of Documents, U.S. Government Print Office, Washington, DC 20402

📖 *Low Pressure Boilers*, 1994, F. M. Steingress, American Technical Publishers, 1155 West 175th Street, Homewood, IL 60403

📖 *System Design Manual - Part III*, Piping Design, Carrier, 1960, Carrier Air Conditioning Company, Carrier Parkway, P.O. Box 4808, Syracuse, NY 13221

📖 *Manual J - Load Calculation for Residential Winter and Summer Air Conditioning*, 1986, 7th edition, Air Conditioning Contractors of America, 1712 New Hampshire Ave., NW, Washington, DC 20009

📖 *Manual N - Load Calculation for Commercial Summer and Winter Air Conditioning*, 1988, 4th edition, Air Conditioning Contractors of America, 1712 New Hampshire Ave., NW, Washington, DC 20009

📖 *Pressure Vessel Code, Section VI - Recommended Rules for the Care and Operation of Heating Boilers*, 1992, American Society of Mechanical Engineers, 22 Law Drive, P.O. Box 2300, Fairfield, NJ 07007-2300

Plumbing Contractor's Exam

📖 *Ohio Plumbing Code, Chapter 4101:2-56-59,* Ohio Administrative Code, July 1995, The Cleveland Plumbing Industry Promotion & Education Fund, 30 Construction Center, 981 Keynote Circle, Brooklyn Heights, OH 44131

📖 *Plumbing Technology: Design and Installation*, 1994, 2nd edition, Lee Smith, Delmar Publishers, P.O. Box 6904, Florence, KY 41022

📖 *Plumbing*, L. V. Ripka, American Technical Publishers, 1155 West 175th Street, Homewood, IL 60403

Recommended Reading for the Contractor's Exams (continued)

Refrigeration Contractor's Exam

📖 *Ohio Basic Mechanical Code*, 1995, BOCA International, 1245 S. Sunbury Road, Suite 100, Westerville, OH 43081-9308

📖 *Code of Federal Regulations - Title 29*, Part 1926 (OSHA), July 1995, Superintendent of Documents, U.S. Government Print Office, Washington, DC 20402

📖 *ANSI/ASHRAE 15-94 - Safety Code for Mechanical Refrigeration*, 1994 ANSI, 11 West 42nd St., New York, NY 10036

📖 *Refrigeration and Air Conditioning*, 1995, 3rd edition, ARI, Prentice Hall, P.O. Box 11071, Des Moines, IA 50336-1071

📖 *System Design Manual - Part III*, Piping Design, Carrier, 1960, Carrier Air Conditioning Company, Carrier Parkway, P.O. Box 4808, Syracuse, NY 13221

📖 *Pipefitters Handbook*, 1967, 3rd edition, Forest Lindsey, Industrial Press, Inc., 200 Madison Avenue, New York, NY 10016

📖 *Trane Air Conditioning Manual*, 1965, Trane Company, 8929 Western Way, Suite #1, Jacksonville, FL 32256

Asbestos Abatement License and Certificates

If you work with asbestos in Ohio, you must be certified and/or licensed by the Ohio Department of Health. Ohio licenses asbestos hazard abatement contractors and certifies these types of asbestos trades:

Asbestos hazard abatement work

Asbestos hazard abatement specialist

Asbestos hazard evaluation specialist

Asbestos hazard project designer

Hazard abatement air-monitoring technician

To get an application for a certificate/license, contact:

Contact

Ohio Department of Health
Asbestos Program
246 North High Street
Columbus, OH 43226
(614) 466-0061
Fax: (614) 752-4157

To qualify for a contractor's license you must:

1. have access to at least one state-approved asbestos disposal site
2. have a worker protection plan approved by EPA, OSHA, and/or Code of Federal Regulations
3. be familiar with all applicable state and federal standard for asbestos hazard abatement projects
4. successfully complete a Department-approved course
5. be capable of complying with all state, EPA, OSHA, Code of Federal Regulation rules
6. register with the Ohio Secretary of State

You must also make sure all employees who come in contact with asbestos or who will be responsible for an asbestos hazard abatement project have fulfilled items 3, 4, and 5 above.

The contractor's license will cost you $500 and it's good for a year.

Here are the qualifications and fees you need to get a certificate in each of the other asbestos work trades:

Trade	Approved training course required?	Exam required?	Fee
Abatement specialist	yes	yes	$125
Evaluation specialist	yes	yes	$125
Abatement project designer	yes	yes	$125
Abatement worker	yes	yes	$25
Abatement air-monitoring technician	yes	no	$75

Certification is good for one year.

Department of Transportation (DOT)

To get an application to prequalify to bid on Department of Transportation work, contact:

Contact

Ohio Department of Transportation
Contractor Qualifications
25 South Front Street, Room 402
P.O. Box 899
Columbus, OH 43216-0899
(614) 466-2823
Fax: (614) 728-2078

Ohio: Department of Transportation (DOT)

The application will ask you which types of work you want to bid on. The types are:

General highway   Fabricated structures

Paving   Painting

Grading   Special

Drainage structures   Roadside improvement

Bridges

The Department will send you a Confidential Financial Statement and Experience Questionnaire to fill in. When you return this information to them, they'll evaluate your company using these percentages:

Factor	Maximum percent
Organization	20
Plant and equipment	20
Construction experience	20
Credit record	15
Performance	25

Your prequalification will be good for one year. However the end date can't be more than four months beyond the last day of your fiscal year.

Out-of-State Corporations

If your business isn't incorporated in Ohio, you must be authorized by the Ohio Secretary of State to do business in the state. You can contact them at:

Contact

Ohio Secretary of the State
30 East Broad Street, 14th Floor
Columbus, OH 43266-0418
(614) 466-3910
Fax: (614) 466-2892

Oklahoma

Oklahoma doesn't license resident construction contractors, except in the electrical, mechanical and plumbing trades. But there are some special requirements for nonresident contractors.

If you're not a resident Oklahoma contractor, you must post a bond equal to three time the tax liability (or 10%) of any contract you get in the state. You also need an employer identification number from the Oklahoma Tax Commission and the Oklahoma Employment Security Commission. To get applications for these items, contact:

Contact

Oklahoma Tax Commission
2501 Lincoln Blvd.
Oklahoma City, OK 73194
(405) 521-4437

Every month the Commission gives workshops (two and one-half hours each) at several locations to help you file the items you need to do business in the state.

Electrician's Licenses

You need a contractor or journeyman license to do electrical work in Oklahoma. To apply for a license, contact:

Contact

Oklahoma State Department of Health
Occupational Licensing Service
Electrical Division
1000 N. E. 10th Street
Oklahoma City, OK 73117-1299
(405) 271-5217

Electrical Contractor's License

You need to give proof of workers' compensation insurance, a $5,000 license bond, and $50,000 liability insurance to apply for a contractor's license. The Division will review your application and if you're eligible they'll send you information on the exam they require. To qualify for the electrical contractor's exam, you must have five years of electrical work experience. Two of the five years must be in commercial/industrial work. You can use up to two and one-half years of Division-approved education to fulfill the experience requirement.

The contractor's exam is open book and lasts about six hours. It has five parts:

- multiple choice questions on the *National Electrical Code*
- practical demo where you connect three- and four-way switches, motor controls, identify electrical tools, fitting, bending conduit
- calculations to design and estimate costs of electrical systems
- business and law
- transformer connections

Electrical contractor's license fees: The electrical contractor exam costs $50 and the license will also cost you $50. A license is good for one year but all licenses expire on June 30 each year.

Journeyman Electrician's License

To qualify for the journeyman's exam you must have four years of electrical work experience. Two of the four years must be in commercial/industrial work. You can use up to two years of Division-approved education to fulfill the experience requirement.

The journeyman's exam is also open book and lasts about six hours. It's on the same items as the contractor's exam, except it doesn't have a business and law section.

Journeyman's electrician's license fees: The journeyman exam costs $25 and the license is $15. A license is good for one year but all licenses expire on June 30 each year.

If you can prove you have a valid license in a state with qualifications at least equal to Oklahoma's, the Division may issue you an Oklahoma license by reciprocity. You still need to pay the fee.

Mechanical Contractor and Journeyman Licenses

You need a license to do mechanical contracting work in Oklahoma. To apply for a license, contact:

Oklahoma State Department of Health
Occupational Licensing Service
Mechanical Division
1000 N. E. 10th Street
Oklahoma City, OK 73117-1299
(405) 271-5217

The Division issues mechanical contractor and journeyman licenses for these trades:

HVAC unlimited (over 25 tons and 500,000 Btu)	Natural gas piping
HVAC limited (under 25 tons and 500,000 Btu or less)	Sheet metal
Refrigeration	Process piping

You need to give proof of workers' compensation insurance, a $5,000 license bond, and $50,000 liability insurance to apply for a contractor's license.

You have to pass exams to get any mechanical license. To qualify for a contractor's exam, you must have four years of work experience. To qualify for a journeyman's exam you must have three years of work experience.

Mechanical Contractor's License

For any contractor license you have to pass a business exam and a rules exam. The business exam has 25 multiple choice questions and lasts 30 minutes. It has math questions on profits, overhead and discounts that you use to operate a business. It also has questions on lien laws, workers' compensation, taxes, and unemployment compensation. The rules exam is open book with 25 multiple choice questions. It also lasts 30 minutes. All of the rules exam questions come from the Mechanical Licensing Act, Mechanical Industry Rules and Regulations, and Fee and Fine Regulations. You can get these publications from the Division. For any journeyman license, you need to pass the rules exam but not the business exam.

Mechanical contractor's license fees: The mechanical contractor's exam and license costs $100. A license is good for one year but all licenses expire on June 30 each year.

Mechanical Trade Licenses

You'll also have to pass trade exams for a license. Here's information on the exams for each trade.

The HVAC exam has five parts. Here's a summary of which of the five parts you have to take for each license and the time allowed for each part:

	Part number and time allowed				
	Part 1 **60 min.**	**Part 2** **30 min.**	**Part 3** **30 min.**	**Part 4** **30 min.**	**Part 5** **30 min.**
Unlimited contractor	▨	▨	▨	▨	▨
Limited contractor	▨	▨	▨		
Unlimited journeyman	▨	▨	▨		
Limited journeyman	▨				

The *refrigeration* exam has two parts. Part 1 has 50 multiple choice questions and lasts 60 minutes. Part 2 has 25 multiple choice questions and lasts 30 minutes. For a contractor license you need to pass both parts. For the journeyman you need Part 2.

The *natural gas piping* exam has two parts. Part 1 has 15 multiple choice questions and lasts 20 minutes. Part 2 has 20 multiple choice questions and lasts 30 minutes. For a contractor license you need to pass both parts. For the journeyman you need Part 2. The 1996 *International Mechanical Code* is the reference for the natural gas piping exam.

The *sheet metal* exam has three parts. Part 1 has 35 multiple choice questions and lasts 45 minutes. Part 2 has 20 multiple choice questions and lasts 30 minutes. Part 3 has 20 questions from a duct layout. For a contractor license you need to pass all parts. For the journeyman you need Part 1.

The *process piping* exam has two parts. Part 1 has 25 multiple choice questions and lasts 30 minutes. Part 2 has 20 multiple choice questions and lasts 30 minutes. For a contractor license you need to pass both parts. For the journeyman you need Part 2.

Fee

Journeyman trade license fees: The journeyman trade exam and license costs $50. A license is good for one year but all licenses expire on June 30 each year.

Recommended Reading for the Mechanical Contractor's Exam

- *Mechanics and Material Lien Law*
- *Handbook of Worker's Compensation*
- *Employer's Tax Guide, Circular E*
- *Employer's Guide About Unemployment Compensation*

HVAC Exam

- *1996 International Mechanical Code*
- *Refrigeration and Air Conditioning Technology*, 3rd edition, W. D. Whitman and W. M. Johnson
- *SMACNA Sheet Metal Symbols*
- *Code of Federal Regulations - Title 29, Part 1926 (OSHA)*, July 1995, Superintendent of Documents, U.S. Government Printing Office, Washington, DC 20402

Recommended Reading for the Mechanical Contractor's Exam (continued)

Refrigeration Exam

📖 *1996 International Mechanical Code*

📖 *Refrigeration and Air Conditioning Technology,* 3rd edition, W. D. Whitman and W. M. Johnson

Sheet Metal Exam

📖 *1996 International Mechanical Code*

📖 *Refrigeration and Air Conditioning Technology,* 3rd edition, W. D. Whitman and W. M. Johnson

📖 *SMACNA Sheet Metal Symbols*

📖 *Code of Federal Regulations - Title 29, Part 1926 (OSHA),* July 1995, Superintendent of Documents, U.S. Government Printing Office, Washington, DC 20402

Process Piping Exam

📖 *1996 International Mechanical Code*

📖 *The Pipe Fitter's and Pipe Welder's Handbook,* revised edition, T. W. Frankland

📖 *Modern Methods of Pipe Fabrication,* 9th edition, S. D. Bowman

📖 *Pipefitters Handbook,* 1967, 3rd edition, Forest Lindsey, Industrial Press, Inc., 200 Madison Avenue, New York, NY 10016

📖 *Code of Federal Regulations - Title 29, Part 1926 (OSHA),* July 1995, Superintendent of Documents, U.S. Government Printing Office, Washington, DC 20402

Plumber's Licenses

You need a license to do most plumbing work in Oklahoma. To apply for a license, contact:

Contact

Oklahoma State Department of Health
Occupational Licensing Service
Plumbing Division
1000 N. E. 10th Street
Oklahoma City, OK 73117-1299
(405) 271-5217

The Division issues plumbing contractor and journeyman licenses. You need to post a $5,000 bond to apply for a contractor license.

You have to pass an exam to get either license. To qualify for the plumbing contractor's exam, you must have four years of plumbing work experience. To qualify for the journeyman's exam you must have three years of plumbing work experience. The exam has written questions, drawings and/or charts, and a practical shop section.

If you can prove you have a valid license in a state with qualifications at least equal to Oklahoma's, the Division may issue you an Oklahoma license by reciprocity. You still need to pay the fee.

Fee

Plumber's license fees: The plumbing contractor exam and license costs $150. The journeyman exam and license costs $55. A license is good for one year but all licenses expire on June 30 each year.

Recommended Reading for the Plumbing Exam

📖 *Plumbing License Law of 1955 and Rules and Regulations Governing Plumbers,* available from the Department of Health

📖 *International Plumbing Code - 1995,* available for $35 from:
OPIA
P.O. Box 2951
Edmond, OK 73083-2951

Department of Transportation (DOT)

Except for right-of-way clearance projects, county bridge projects, and a few special projects, you have to be prequalified to bid on Oklahoma Department of Transportation projects. To get an application, contact:

Contact

Department of Transportation
200 21st Street
Oklahoma City, OK 73105
(405) 521-2561
Fax: (405) 522-1035

Some of the details the Department will ask you about are your organization, personnel, financial condition, equipment, and experience. You'll be asked what type of work you want to be qualified for. Here are the types of work the Department uses:

Grading

Minor structure

Paving

Bridge

Traffic control, lighting

Other

You'll also be asked to appoint an Oklahoma resident who will be your agent for receipt of legal process. You must register with the Oklahoma Secretary of State as well.

Prequalification is good for up to 24 months from the date of final acceptance of your last project.

Out-of-State Corporations

An out-of-state corporation needs to register with the Oklahoma Secretary of State to do business in the state. For information, contact:

Contact

Office of the Secretary of State
State Capitol Building, Room 101
Oklahoma City, OK 73105
(405) 521-3911
Fax: (405) 521-3771

Oregon

If you're paid for any construction activity, you need to register with the Oregon Construction Contractors Board. To get an application, contact:

Contact

Construction Contractors Board
700 Summer Street NE, Suite 300
P.O. Box 14140
Salem, OR 97309-5052
(503) 378-4621 ext. 4900
Fax: (503) 373-2007

Before you can register with the Board, you must identify one member of your business as a Responsible Managing Individual (RMI). Usually the RMI is an owner, partner or corporate officer of the business. The RMI must complete a four-hour course on:

- Construction Contractors Board registration and application requirements
- Oregon employer requirements
- tax accounting
- Oregon building codes
- occupational safety
- environmental protection
- contract law
- Oregon lien law

Then the RMI must complete four courses (12 hours) from this list:

- basic contract law
- Oregon's lien law
- basic accounting
- codes and licenses
- OR-OSHA practices
- job site safety and security
- current employee issues
- construction finance
- government construction work

- project management
- dispute prevention
- environmental protect requirements
- cultural diversity and construction
- estimating
- customer service

There are about a dozen places in Oregon where you can take some of these courses, including:

Name	Phone number
Acorn Legal Resources, Inc.	(800)-597-7161
American Institute of Construction Management	(800) 644-2426
Construction Contracting Academy	(800) 937-2242
Central Oregon Community College Business Development Center	(541) 383-7290
Lane Community College Small Business Development Center	(541) 726-2255
Marion-Polk Building Industry Association	(503) 399-1500
Owner Builder School	(503) 631-8007
Oregon Contractors Education Services	(800) 859-0325
Oregon Building Industry Association	(503) 378-9066
Portland Community College Small Business Development Center	(503) 978-5080
Portland Home Builders Association	(503) 684-1880
Rogue Community College Small Business Development Center	(541) 471-3515
Vo-Tech Institute	(800) 650-0508

There are some special cases that let you skip these requirements. If, within the last seven years before applying to the Oregon Board, the RMI has been licensed in a state that registers or licenses its contractors, you may be exempt from some of the courses. But you'll still have to take the four-hour course on Oregon laws and business practices. Check with the Board for these qualifications. Here's a list of the states the Board accepts experience in:

Alabama	Kansas	New Mexico
Alaska	Louisiana	North Carolina
Arkansas	Maryland	North Dakota
Arizona	Massachusetts	Rhode Island
California	Michigan	South Carolina
Connecticut	Minnesota	Tennessee
Delaware	Missouri	Utah
Florida	Montana	Virginia
Georgia	Nebraska	Washington
Hawaii	Nevada	Washington DC
Iowa	New Jersey	

You'll also need to qualify as an independent contractor to register with the Board.

Here are the types of licenses the Board issues, bond and insurance requirements, and one-year registration fees:

License type	Bond	Insurance	Fee
General Contractor - all structures	$10,000	$500,000	$100
General Contractor - residential only	$10,000	$100,000	$80
Special Contractor - all	$5,000	$500,000	$100
Special Contractor - residential only	$5,000	$100,000	$80
Limited Contractor (under $30,000 gross/year)	$2,000	$100,000	$100

Asbestos Abatement License and Certifications

If you do asbestos abatement work in Oregon, you must be licensed or certified by the Department of Environmental Quality. The Department licenses contractors and certifies supervisors or workers. To get an application, contact:

Contact

Oregon DEQ Asbestos Section
2020 SW Fourth, Suite 400
Portland, OR 97201-4987
(503) 229-5982
Fax: (503) 229-6957
(800) 452-4011 (Oregon only)

To get a contractor's license, you must:

- be or employ a certified supervisor
- read, understand, and comply with the applicable Oregon and federal rules on asbestos abatement
- prove workers' compensation coverage (not required for independent contractors)
- register with the Oregon Construction Contractors Board
- register with the Oregon Business Registry (Oregon Secretary of State)

Fee

Asbestos abatement license fees: The license will cost $1,000 and it's good for one year.

To be certified as a supervisor you must successfully complete the supervisor level training and exam approved by the Department. You must also have one of the following:

- three months experience in asbestos abatement and five separate asbestos abatement projects

- a current Oregon Worker Certification and six months experience as supervisor in general construction, general environmental, or facility maintenance of an asbestos abatement project

Asbestos abatement certificate fee: If you have a certificate, training, and/or experience in another state, Oregon may accept them. The certificate will cost $65 and it's good for one year.

To be certified as a worker you must successfully complete the worker level training and exam approved by the Department. The certificate costs $45 and it's good for one year.

Electrician's Licenses

To do electrical work in Oregon you need a license. You also need to register with the Oregon Construction Contractors Board. To find out about the Construction Contractors Board, see the information at the beginning of the Oregon section. To get an application for a license, contact:

Department of Consumer & Business Services
Building Codes Division
1535 Edgewater NW
Salem, OR 97310
(503) 373-1268
Fax: (503) 378-2322

The Division issues eight electrical contractor licenses. Here they are, with the license fee for each:

Contractor license	Fee
Elevator	$195
Limited energy electrical	$125
Limited maintenance specialty	$25
Limited pump installation	$25
Electrical	$125
Limited sign	$125
Limited maintenance specialty HVAC/R	$25
Restricted energy electrical HVAC	$125

You don't have to take an exam to get an electrical contractor's license. However you do need to employ an appropriately-licensed person in your company to sign for electrical permits. There are also these requirements for four of the contractor's licenses:

Contractor license	Requirements
Limited maintenance specialty	2,000 hours experience with appliances, fluorescent ballast, or similar equipment
Limited maintenance specialty HVAC/R	4,000 hours experience in HAVC/R work
Limited pump installation	2,000 hours experience with pump equipment or irrigation water systems on residential property
Restricted energy HVAC	4,000 hours experience with HVAC work

You need to be registered with the Construction Contractors Board to get any of these licenses. All electrical contractor licenses expire on October 1 each year.

The Division also issues these electrical licenses:

General Supervisor
Limited Supervisor Manufacturing Plant
General Journeyman
Limited Journeyman Manufacturing Plant
Limited Journeyman Limited Energy
Restricted Energy Technician
Limited Residential
Limited Maintenance Electrician
Limited Journeyman Sign
Limited Journeyman Stage
Limited Journeyman Railroad
Limited Maintenance Mobile Home/RV
Limited Journeyman Elevator
Limited Building Maintenance

For any of these licenses, you'll have to satisfy the Division requirements for the license and pass a written exam (except there's no exam for the Limited Journeyman Elevator license). The Division will review your application and notify you if you're eligible to take the exam. Here's information on Division requirements and the exam for each license.

For the *General Supervisor* license you need four years experience with a General Journeyman license or Division-approved equivalent experience of eight years (16,000 hours) with 2,000 each in residential, individual, and commercial work. The exam is open book and has two parts — Part I with 75 multiple choice questions (four hours) and Part II with two calculation questions (two hours).

For the *General Journeyman* license you need to complete a four-year approved apprenticeship program or Division-approved equivalent experience. The exam is open book with 50 multiple choice questions and lasts three hours. You can get a reciprocal license from the Division if you have a current General Journeyman license in Alaska, Idaho, Maine, Massachusetts, North Dakota, Rhode Island, Washington or Wyoming. You must have earned your license by taking the state exam. The fee for this license is $100 and it's good for three years.

For the *Limited Journeyman Limited Energy* license you need to complete an approved apprenticeship program or Division-approved equivalent experience. The three-hour exam is open book with 50 multiple choice questions. You can get a reciprocal license from the Division if you meet these requirements:

- you have a current Limited Journeyman Limited Energy license in Washington
- you were a Washington resident when you took the exam for the license
- you can verify two years of work experience as a Limited Journeyman Limited Energy electrician after you got the Washington license

For the *Limited Residential* license you need to complete a two-year approved apprenticeship program or Division-approved equivalent experience. The exam is open book with 50 multiple choice questions and lasts three hours.

For the *Limited Journeyman Sign* license you need to complete a four-year approved apprenticeship program or Division-approved equivalent experience. The three-hour exam is open book with 50 multiple choice questions. You can get a reciprocal license from the Division if you meet these requirements:

- you have a current Limited Journeyman Sign license in Washington
- you were a Washington resident when you took the exam for the license
- you can verify two years of work experience as a Limited Journeyman Sign electrician after you got the Washington license

For the *Limited Journeyman Railroad* license you need to complete a four-year approved apprenticeship program or Division-approved equivalent experience. The exam is open book with 25 multiple choice questions and lasts three hours.

For the *Limited Journeyman Elevator* license you need to hold a National Elevator Industry Educational Program Certificate. There's no exam for this license.

For the *Limited Supervisor Manufacturing Plant* license you need at least four years of work experience with a Limited Journeyman Manufacturing Plant license. The three-hour exam is open book with 50 multiple choice questions.

For the *Limited Journeyman Manufacturing Plant* license you need to complete an approved apprenticeship program or Division-approved equivalent experience. The three-hour exam is open book with 50 multiple choice questions.

For the *Restricted Energy Technician* license you need to complete a two-year approved apprenticeship program or Division-approved equivalent experience. The three-hour exam is open book with 25 multiple choice questions.

For the *Limited Maintenance Electrician* license you need at least two years (4,000 hours) experience maintaining industrial plant equipment. The three-hour exam is open book with 25 multiple choice questions.

For the *Limited Journeyman Stage* license you need to complete a four-year approved apprenticeship program or Division-approved equivalent experience. The exam is open book with 50 multiple choice questions and lasts three hours.

For the *Limited Maintenance Mobile Home/RV* license you need at least two years (4,000 hours) experience with a manufacturer or a licensed Limited Maintenance Mobile Home/RV electrician. The three-hour exam is open book with 25 multiple choice questions.

For the *Limited Building Maintenance* license you need at least one year (2,000 hours) on-the-job training or experience in commercial building maintenance or Division-approved equivalent experience. The three-hour exam is open book with 25 multiple choice questions.

Electrician's license fees: It will cost you $10 to file an application for any license (except the Limited Journeyman Elevator license application, which is $15). Here's a summary of the licenses and the fee for each one:

License	Fee
General Supervisor	$100
Limited Supervisor Manufacturing Plant	$100
General Journeyman	$100
Limited Journeyman Manufacturing Plant	$100
Limited Journeyman Limited Energy	$50
Restricted Energy Technician	$50
Limited Residential	$100
Limited Maintenance Electrician	$100
Limited Journeyman Sign	$50
Limited Journeyman Stage	$100
Limited Journeyman Railroad	$100
Limited Maintenance Mobile Home/RV	$100
Limited Journeyman Elevator	$50
Limited Building Maintenance	$100

Licenses are good for three years and the fee will be prorated over that period.

Recommended Reading for the Electrical Exams

📖 *National Electrical Code,* including Wire Tables (current edition)

📖 *Oregon Specialty Code* (current edition)

📖 *NEC Handbook,* 1996, NFPA, Batterymarch

📖 *American Electricians Handbook,* 1996, 13th edition, Croft/Summers, McGraw-Hill Inc., Box 543, Blacklick, OH 43004-0543

📖 *Ferm's Fast Finder Index,* Olaf G. Ferm, Ferms Finder Index Company

📖 *Tom Henry Key Word Index*

📖 *Ugly's Fast Finder Index*

Plumber's Licenses

To do plumbing work in Oregon you need a license. You also need to register with the Oregon Construction Contractors Board. And you need to register your business with the Oregon Department of Consumer & Business Services. To find out about the Construction Contractors Board, check the information at the beginning of the Oregon section. To register your business with Consumer & Business Services you need to fill out a Boiler/Plumbing Business Registration Application. You can get this from:

Contact

Department of Consumer & Business Services
Building Codes Division
1535 Edgewater NW
Salem, OR 97310
(503) 373-1268
Fax: (503) 378-2322

This registration costs $150 and it's good for one year.

The Division also issues three types of plumbing licenses — journeyman, one- and two-family dwelling residential water heater installer, and water treatment installer. To get an application, contact the Building Codes Division.

The Division will review your application and if you're eligible they'll send you information on the exam they require. For the journeyman license you need a high school diploma or equivalent and four years of plumbing experience or completion of an approved apprenticeship program. For the Water Treatment Installer license you need a high school diploma or equivalent and 18 months of water treatment installation experience or a plumbing apprentice program or 3,000 hours of work experience. The Division may accept military experience if it's relevant and properly documented.

The exam is closed book with 100 questions and lasts three hours. It covers:
- drains and vents/sewers
- water supply/backflow prevention
- general code knowledge
- installation practices/methods materials
- special wastes/roof drains
- fixtures/trim
- excavation
- inspection/test/permits

After you pass the exam with a score of 75% or better, you need to pass a practical exam which tests your mechanical skills on fabricating and assembling pipes, fixtures, and other apparatus.

If you have a valid plumbing license from any of the following states, you may be able to get a journeyman license without taking the exam:

Alaska	Kansas	North Dakota
Arkansas	Louisiana	Rhode Island
Colorado	Maine*	South Dakota
Connecticut	Maryland	Texas*
Delaware	Massachusetts	Virginia
Hawaii	Michigan	Wisconsin**
Illinois	Minnesota	

*Masters license only accepted
**Apprenticeship and journeyman license only accepted

If you have an active journeyman license from Idaho or Washington that you got by taking an exam, you can apply for a reciprocal journeyman license in Oregon.

For the 1 & 2 Family Dwelling Residential Water Heater Installer license you need to go to an approved training class for the trade and pass a Limited Specialty Plumbing 1 & 2 Family Dwelling Residential Water Heater Installer exam.

Here are the licenses and what they cost:

License	Fee
Journeyman	$35
Reciprocal Journeyman	$95
Water Treatment Installer	$35
1 & 2 Family Dwelling Residential Water Heater Installer	$25

Recommended Reading for the Plumbing Exams

📖 *Oregon State Plumbing Specialty Code*, 1996 edition

📖 *Oregon State One and Two Family Dwelling Specialty Code*, 1995 edition

---

Department of Transportation (DOT)

To bid as a prime contractor on Oregon Department of Transportation jobs, you must be prequalified. Suppliers and subcontractors don't need to be prequalified. To get an application, contact:

Contact

Department of Transportation
307 Transportation Building
Salem, OR 97310
(503) 986-3877

The application will ask you about your company organization, equipment, experience, and what licenses you hold. You will also need to fill in a table telling the maximum dollar amount you're capable of performing, maximum dollar amount you're qualified for in other states, and years of experience for these classes of work:

Land clearing	Sewer construction
Earthwork and drainage	Building construction, alteration, repair
Aggregate crushing and bases	Electrical wiring
Asphalt concrete pavement and oiling	Painting and decorating
Portland cement pavement	Demolition
Bridges and grade separations	Heating
Painting bridges and grade separations	Plumbing
Signing	Roofing
Miscellaneous highway appurtenances	Air conditioning
Traffic signals	Drainage
Illumination	Sheet metal work
Buildings	Municipal street construction
Landscaping	Well drilling
Water lines, reservoirs, tanks	Other

Fee

Department of Transportation application fee: It costs $100 to file the application for prequalification. Once you're prequalified, the DOT will notify you when you need to be requalified.

Out-of-State Corporations

Out-of-state corporations must register with the Oregon Secretary of State. For information, contact:

Contact

Corporations Division
Oregon Secretary of State
255 Capitol Street, NE, Suite 151
Salem, OR 97310-1327
(503) 986-2200
Fax: (503) 378-4381

Pennsylvania

The only special requirements for construction contractors in Pennsylvania are in the areas of public works and the Department of Transportation.

Public Works

To bid on public works projects in Pennsylvania you need to get your name on the Department of General Services Contractors List. To do this, ask for a Contractor's Questionnaire form from:

Department of General Services
Commonwealth of Pennsylvania
18th and Herr Streets, Room 103
Harrisburg, PA 17125
(717) 783-7610
Fax (717) 772-3399

The questionnaire will ask you about what types of work you want to do, the value of projects you may bid on, minority status of your business, and which counties of Pennsylvania you want to work in.

You'll need to pay $50 to file the questionnaire. It's good for one year.

Department of Transportation (DOT)

To bid on Pennsylvania Department of Transportation projects you have to be prequalified by the Department. To get an application for prequalification, contact:

Pennsylvania Department of Transportation
Prequalification Office
Forum Place, 7th Floor
555 Walnut St.
Harrisburg, PA 17101-1900
(717) 787-7032

Some of the details the Department will ask you about are your organization, personnel, financial condition, equipment, and experience. The Department will also check if you have any tax liabilities, suspensions from any state agency, and deficient performance on any previous state contract.

You'll also be asked what type of work you want to be qualified for. Here are the types of work the Department uses:

Earthwork:
　Clearing, grubbing
　Excavating, grading
　Building demolition

Base course:
　P.C.C. base course
　Flexible base course

Pavement:
　Bituminous
　Bituminous patching, manual repair
　Rigid pavement
　Rigid patching, repair

Incidental construction:
　Drainage, water main, storm sewer
　Curbs, sidewalks, inlets, manholes
　Guide rail, steel median barrier, fences
　Slabjacking, subsealing

Roadside development:
　Landscaping
　Rest area structures, buildings

Traffic accommodations, control:
　Pavement markings
　Sign placement
　Highway/sign lighting, signal control
　Sign structures

Structures:
　Cement concrete structures
　Repair, rehabilitation of structures
　Erection

Pile driving

Metal plate, concrete box culverts, short span bridges

Modified concrete deck overlays

Bridge removal

Steeling painting

Miscellaneous:

Electrical

Plumbing

Other

Prequalification is good for 18 months from the date of the balance sheet you send in with your prequalification application.

Rhode Island

If you build, repair, or remodel one- to four-family dwellings in Rhode Island, you must register with the Contractors' Registration Board. To get an application, contact:

Department of Administration
Contractors' Registration Board
One Capitol Hill
Providence, RI 02908-5859
(401) 277-1268
Fax: (401) 277-2599

There's no exam for this registration. However you must supply all of this information on your application:

- complete information on the ownership and organization of your company
- proof of liability insurance for at least $300,000
- any current or previous registrations you have as a contractor
- your specialty trades
- years of experience in construction
- workers' compensation insurance account number
- unemployment insurance account number
- federal employer identification number

It will cost you $60 to register with the Board. Registration is good for one year.

Asbestos Abatement Licenses

You need a license to do asbestos abatement work in Rhode Island. To get an application, contact:

Department of Health
Office of Occupational & Radiological Health
206 Cannon Building
Providence, RI 02908-5097
(401) 277-3601
Fax: (401) 277-2456

The Department issues asbestos contractor, asbestos abatement site supervisor, and asbestos abatement worker licenses.

To qualify for a contractor's license you must:

1. Have a worker protection and medical monitoring plan approved by EPA, OSHA, and/or Code of Federal Regulations

2. Be familiar and comply with all applicable state and federal standards for asbestos abatement projects

3. Provide company standard operating training for all employees before starting a job

4. Be capable of complying with all state, EPA, OSHA, and Code of Federal Regulation rules

5. Provide a list of personnel who have completed Department-approved CPR and basic first aid training

6. Provide a list of all personnel (to include at least one licensed site supervisor) and asbestos equipment

7. Provide list of states you are licensed in for asbestos work

8. Provide list of any citations received in last two years for asbestos work

9. Provide proof that all employees have been licensed by the Department

You must also make sure all employees who come in contact with asbestos or who will be responsible for an asbestos abatement project have fulfilled items 2, 3, and 4 above.

To get an asbestos site supervisor license, you must be licensed together with an asbestos contractor. You can't get an independent site supervisor license. You need to take a 40-hour site supervisor training course to apply for a license.

The same qualifications for licensing apply to an asbestos worker license. The license is with an asbestos contractor. The asbestos worker training course lasts 32 hours.

Asbestos abatement license fees: A contractor license will cost you $1,500, a supervisor license is $40, and a worker license is $20. The contractor and supervisor licenses are good for two years. The worker license is good for one year.

Electrician's Licenses

You need a license to do electrical work in Rhode Island. To get an application, contact:

Division of Professional Regulation
Board of Examiners of Electricians
610 Manton Avenue
Providence, RI 02909
(401) 457-1860
Fax: (401) 457-1868

The Board issues ten types of licenses. Here they are, with the work experience requirement and license fee for each one:

Electrical contractor

Required training or experience: 12,000 hours plus journeyman license for two years

Fee: $100

Electrical journeyman

Required training or experience: 8,000 hours as a registered apprentice

Fee: $30

Limited premises

Fee: $100

Elevator journeyman

Required training or experience: 8,000 hours as a registered apprentice plus National Elevator Industry Education Program

Fee: $30

Oil burner contractor

Required training or experience: 12,000 hours plus oil burner man license for two years

Fee: $100

Oil burner man

Required training or experience: 8,000 hours as a registered apprentice

Fee: $30

Fire alarm contractor

Required training or experience: 12,000 hours plus fire alarm installer license for two years

Fee: $100

Fire alarm installer

Required training or experience: 8,000 hours as a registered apprentice

Fee: $30

Electrical sign contractor

Required training or experience: 12,000 hours plus electrical sign installer license for two years

Fee: $100

Electrical sign installer

Required training or experience: 8,000 hours as a registered apprentice

Fee: $30

Except for the limited premises license, you must pass an exam to get a license. The exams are all based on the *National Electrical Code*. If you have a contractor or journeyman license in another state, you may take the Rhode Island journeyman exam. If you have a contractor license in another state, a Rhode Island journeyman license for two years, and 12,000 hours of work experience, you may take the Rhode Island contractor exam.

Electrician's license fees: It'll cost you $30 to take any exam.

Your first license will expire on your birthday. When your renew your license, it'll be good for two years.

Mechanical Work Licenses

You need a license to do mechanical work in Rhode Island. To get an application, contact:

Division of Professional Regulation
610 Manton Avenue
Providence, RI 02909
(401) 457-1860
Fax: (401) 457-1868

The Division issues 18 types of mechanical licenses. Here's a list of them, with the work experience requirement and license fee for each one:

Master mechanical contractor

Required training or experience: 10 years experience as master pipefitter I and 10 years as master refrigeration contractor I

Fee: $100

Master pipefitter I

Required training or experience: One year as journeyman pipefitter I or one year as master pipefitter II or valid out-of-state master pipefitter I license or six years approved out-of-state work experience

Fee: $100

Master pipefitter II

Required training or experience: One year as journeyman pipefitter II or valid out-of-state master pipefitter II license or six years approved out-of-state work experience

Fee: $40

Journeyman pipefitter

Required training or experience: Complete approved apprenticeship program or valid out-of-state journeyman pipefitter I license or five years approved out-of-state work experience

Fee: $30

Journeyman pipefitter II

Required training or experience: Complete approved apprenticeship program or valid out-of-state journeyman pipefitter II license or five years approved out-of-state work experience

Fee: $25

Master refrigeration contractor I

Required training or experience: One year as journeyman refrigeration I or one year as master refrigeration II or valid out-of-state master refrigeration I license or six years approved out-of-state work experience

Fee: $100

Master refrigeration contractor II

Required training or experience: One year as journeyman refrigeration II or valid out-of-state master refrigeration II license or six years approved out-of-state work experience

Fee: $40

Journeyman refrigeration contractor I

Required training or experience: Complete approved apprenticeship program or valid out-of-state journeyman refrigeration I license or five years approved out-of-state work experience

Fee: $30

Journeyman refrigeration contractor II

Required training or experience: Complete approved apprenticeship program or valid out-of-state journeyman refrigeration II license or five years approved out-of-state work experience

Fee: $25

Master sprinklerfitter contractor I

Required training or experience: One year as journeyman sprinklerfitter I or valid out-of-state master sprinklerfitter I license or six years approved out-of-state work experience

Fee: $100

Journeyman sprinklerfitter contractor I

Required training or experience: Complete approved apprenticeship program or valid out-of-state journeyman sprinklerfitter I license or five years approved out-of-state work experience

Fee: $30

The Division also issues limited mechanical licenses. These limited trades usually require a two-year apprenticeship program. The license fee for each of them is $25 (except the journeyman oil burner serviceman II, which costs $60). Here are the limited mechanical licenses:

Fire suppression journeyman II

Welding journeyman II

Gas station journeyman II

Oil burner serviceman journeyman II

Gas journeyman II

Oil burner journeyman II

Sheet metal journeyman II

Mechanical work license fees: Except for the master mechanical contractor license, you must pass an exam to get any license. It will cost you $30 to take any exam.

To get any of these licenses, you must also post a $3,000 bond:

Master mechanical contractor

Master pipefitter I

Master pipefitter II

Master refrigeration contractor I

Master refrigeration contractor II

Your first license will expire on your birthday. When your renew your license it will be good for two years.

Plumber's Licenses

You need a license to do plumbing work in Rhode Island. To get an application, contact:

Department of Health
Division of Professional Regulation
3 Capitol Hill, Room 104
Providence, RI 02908
(401) 277-2827
Fax: (401) 277-1272

The Department issues master and journeyman plumber licenses. You have to pass an exam for either license. For a master license you also need to have one year of experience as a licensed journeyman. If you have a master license in another state that you got by passing an exam, you can use it to qualify to take the Rhode Island exam.

To get a journeyman license you also need four years of apprentice plumber experience. If you have a plumbing license in another state, you can apply with the Department to use it to qualify to take the Rhode Island journeyman exam. The exam is based on the *1995 International BOCA Code*, Building Officials & Code Administrators International, 4051 Flossmoor Road, Country Club Hills, IL 60478-5795.

Plumber's license fees: It costs $100 nonrefundable to apply for a master plumber license. The journeyman application fee is $50. A master license costs $60 and a journeyman license, $25. Licenses are good for one year but all expire on June 30th each year.

Department of Transportation (DOT)

The Rhode Island Department of Transportation doesn't require bidders on projects to be prequalified by the Department. To get information on Department of Transportation contracts, contact:

Department of Transportation
Contracts Division
2 Capitol Hill, Room 331
Providence, RI 02903
(401) 277-2495

Out-of-State Corporations

Corporations doing business in Rhode Island must register with the Rhode Island Secretary of State to do business in the state. For information, contact:

Corporations
100 North Main Street
Providence, RI 02903
(401) 277-3040
Fax: (401) 277-1309

South Carolina

To do residential building in South Carolina you must be licensed. To get an application for a license, contact:

Contact

South Carolina Department of Labor, Licensing and Regulation
Residential Builders Commission
3600 Forest Drive, Suite 100
P. O. Box 11329
Columbia, SC 29211-1329
(803) 734-4255
Fax (803) 734-4267

You will have to pass an exam to get your license and you must have one year of work experience under the supervision of a licensed residential builder to take the exam. The exam is on plans, regulations, contracts, and standards.

Fee

Contractor's license fees: It will cost you $100 to file the application. After you pass the exam, you'll need to pay a $100 license fee and submit either a notarized financial statement showing your net worth as $50,000 or more or a surety bond for $15,000. The license expires July 1 each year.

If you have a license from another state whose requirements are comparable to South Carolina's, the Commission may grant you a license without taking the exam. Check with the Commission on this.

Recommended Reading for the Residential Builder's License

📖 *NFPA 70A - National Electrical Code for One and Two Family Dwellings,* National Fire Protection Association, 1 Batterymarch Park, P. O. Box 9101, Quincy, MA 02269

📖 *CABO, One and Two Family Dwelling Code,* Southern Building Code Congress International, Birmingham, AL

General or Mechanical Contractor

To do general or mechanical contracting work in South Carolina you must get a license from the South Carolina Contractors' Licensing Board. The Board issues these types of general contracting licenses:

Building

- building contractor
- general contractor

Highway

- bridges
- roads (grading, paving)
- roads (grading)
- roads (paving)
- incidental

Public utilities

- gas lines
- water and sewer lines
- water and sewer plants
- electrical
  (poles and transportation lines)

Specialties

- communication/instrumentation
- concrete
- heavy
- marine
- masonry
- metal buildings
- nonstructural
- process piping
- railroad
- refrigeration
- roofing
- steel
- swimming pools

Wood-framed structures I, II

The Mechanical licenses they issue are:

- Plumbing
- Electrical
- Heating
- Lightning protection systems
- Air conditioning
- Packaged equipment

Bidder license fees: You also need a bidder's license if you're negotiating a contract with a property owner as a prime contractor. It'll cost you $220 for a general or mechanical license and $240 for a bidder's license. All licenses are good for two years but they expire on December 31 of the second year. To get an application for a license, contact the Board at:

South Carolina Contractors' Licensing Board
South Carolina Department of Labor, Licensing and Regulation
P. O. Box 11329
Columbia, SC 29211-1329
(803) 734-4185

Along with the application, you'll get an open-book, take-home exam with 25 questions on the South Carolina Code of Laws on contracting in the state. You'll also have to take a trade exam which will cost you $50. You have to pass the trade exam for any of the general or mechanical contracting licenses except Highway — incidental, and Specialties — heavy, masonry, nonstructural, railroad, steel and communication/instrumentation.

Building Contractor

The Building contractor exam has two parts:

Part I (open book, 50 questions, three hours, 50% of grade)

Subject	Number of questions
Concrete methods, installation	3 - 5
Concrete materials	4 - 6
Reinforcing steel	5 - 7
Formwork	6 - 8
Structural steel	3 - 5
Masonry	7 - 9
Wood	5 - 7
Foundations	4 - 6
Safety code	1 - 3
Drywall	2 - 4

Part II (open book, 50 questions, three hours, 50% of grade)

Subject	Number of questions
Safety — Code of Federal Regulations	9 - 11
General estimating	6 - 8
Formwork design	9 - 11
Wood design	7 - 9
Estimating — quantity surveying	9 - 11
Business math	4 - 6

If you have a Building license in North Carolina you may be able to get your South Carolina license without taking the exam. Check with the Board on this.

General Contractor

The General contractor exam has two parts:

Part I (open book, 50 questions, three hours, 50% of grade)

Subject	Number of questions
Concrete methods, installation	6 - 8
Concrete materials	5 - 7
Reinforcing steel	3 - 5
Formwork	3 - 5
Structural steel	5 - 7
Masonry	6 - 8
Wood	6 - 8
Foundations	2 - 4
Safety code	2 - 4
Drywall	2 - 4

Part II (open book, 50 questions, three hours, 50% of grade)

Subject	Number of questions
Safety — Code of Federal Regulations	9 - 11
General estimating	6 - 8
Formwork design	9 - 11
Wood design	7 - 9
Estimating — quantity surveying	9 - 11
Business math	4 - 6

Highway Bridges

The Highway Bridges exam is open book, three hours long, with 50 questions. Here are its contents:

Subject	Number of questions
Concrete	8 - 15
Reinforcing steel	4 - 6
Formwork	3 - 5
Structural steel	6 - 10
Rigging	6 - 10
Paving	2 - 5
Safety — Code of Federal Regulations	4 - 6
Estimating	2 - 5
Business math	1 - 3

If you have a Highway license in North Carolina you may be able to get your South Carolina license without taking the exam. Check with the Board on this.

Highway Grading

The Highway Grading exam is open book, three hours long, with 40 questions. Here are its contents:

Subject	Number of questions
General knowledge	14 - 16
Excavating and grading	19 - 21
Safety — Code of Federal Regulations	4 - 6

If you have a Highway or Grading license in North Carolina, you may be able to get your South Carolina license without taking the exam. Check with the Board on this.

Highway Paving

The Highway Paving exam is open book, three hours long, with 50 questions. Here are its contents:

Subject	Number of questions
Compaction and stabilization	6 - 10
Asphalt paving procedures	6 - 10
Asphalt properties and uses	6 - 10
Asphalt emulsions	6 - 10
Asphalt maintenance	6 - 10
Concrete	6 - 10

If you have a Highway license in North Carolina, you may be able to get your South Carolina license without taking the exam. Check with the Board on this.

Public Utilities Gas Lines

The Public Utilities Gas Lines exam is open book, three hours long, with 50 questions:

Subject	Number of questions
Code compliance	5 - 10
Fuel gas systems	15 - 20
Piping	6 - 12
Welding	4 - 6
Safety	4 - 6

Public Utilities Water and Sewer Lines

The Public Utilities Water and Sewer Lines exam is open book, three hours long, with 60 questions:

Subject	Number of questions
Safety — Code of Federal Regulations	3 - 5
General knowledge	3 - 5
Miscellaneous engineering-cut sheet calculations	5 - 7
Underground concrete pipe installation	6 - 8
Underground ductile pipe installation	14 - 18
Underground clay pipe installation	12 - 20
Excavating and grading	2 - 14

If you have a Water and Sewer Lines license in North Carolina, you may be able to get your South Carolina license without taking the exam. Check with the Board on this.

Public Utilities Water and Sewer Plants

The Public Utilities Water and Sewer Plants exam is open book, three hours long, with 50 questions:

Subject	Number of questions
Safety — Code of Federal Regulations	3 - 5
General plant construction	18 - 22
Concrete	5 - 8
Concrete-reinforcing steel	4 - 8
Excavating and grading	4 - 12
Miscellaneous engineering-cut sheet calculations	5 - 7

If you have a Water and Sewer Plants license in North Carolina, you may be able to get your South Carolina license without taking the exam. Check with the Board on this.

Public Utilities Electrical

The Public Utilities Electrical exam is open book, three hours long, with 30 questions:

Subject	Number of questions
Power line construction	23 - 25
Cross arm and transformer problems	5 - 7

Concrete

The Concrete exam is open book, three hours long, with 50 questions:

Subject	Number of questions
Safety — Code of Federal Regulations	5 - 7
Concrete materials	10 - 12
Reinforcing steel installation	3 - 5
Reinforcing steel tools, equipment, fasteners	2 - 4
Formwork design	14 - 16
Concrete methods, installation	10 - 12

Marine

The Marine exam is open book, three hours long, with 50 questions:

Subject	Number of questions
Marine safety	5 - 10
Miscellaneous engineering-pile driving	20 - 30
Concrete	12 - 18
Concrete-reinforcing steel	2 - 5

Process Piping

The Process Piping exam is open book, three hours long, with 60 questions:

Subject	Number of questions
Code compliance	5 - 10
Fuel gas systems — piping	10 - 20
Piping	10 - 20
LPG gas systems	4 - 8
Boilers — steam traps	5 - 10
Process piping (dry cleaning)	3 - 6
Welding	3 - 6
Safety	2 - 5

Refrigeration

The Refrigeration exam is open book, three hours long, with 50 questions:

Subject	Number of questions
Code compliance	5 - 10
HARV general	7 - 12
HARV maintenance	5 - 10
HARV controls	5 - 10
Refrigeration safety	5 - 10
General piping	2 - 5
Piping-dimensioning and design	5 - 10
Safety	0 - 5

If you have a Refrigeration license in Texas, you may be able to get your South Carolina license without taking the exam. Check with the Board on this.

Roofing

The Roofing exam is open book, three hours long, with 50 questions:

Subject	Number of questions
Safety — Code of Federal Regulations	5 - 7
Roofing — materials	6 - 8
Roofing — installation	12 - 14
Roofing — tools, equipment, fasteners	3 - 5
Roofing — substrate	2 - 4
Roofing — definitions	2 - 4
Roofing — quantities	6 - 8
Roofing — code	6 - 8

If you have a Roofing license in North Carolina, you may be able to get your South Carolina license without taking the exam. Check with the Board on this.

Swimming Pool

The Swimming Pool exam is open book, three hours long, with 50 questions:

Subject	Number of questions
Pool — electrical	4 - 8
Gunite	6 - 10
Concrete	4 - 8
Concrete — reinforcing steel	4 - 8

Pool care and operation	10 - 15
Regulation 61-51 — swimming pools	6 - 10
Pool — estimating	2 - 5

Wood Framed Structures I

The Wood Framed Structures I exam is open book, two hours long, with 50 questions:

Subject	Number of questions
Sitework	3 - 5
Concrete	9 - 11
Wood and plastics	13 - 17
Thermal and moisture protection	7 - 9
Doors and windows	2 - 4

Wood Framed Structures II

The Wood Framed Structures II exam is open book, two hours long, with 40 questionst:

Subject	Number of questions
Sitework	2 - 4
Concrete	4 - 6
Wood and plastics	10 - 14
Thermal and moisture protection	7 - 9
Doors and windows	1 - 3

Plumbing I

The Plumbing exam has three parts:

Part I (closed book, 100 questions, two hours, 40% of grade)

Subject	Number of questions
Administration	8 - 12
Basic principles	1 - 3
Definitions	4 - 6
General regulations	10 - 12
Materials	3 - 5
Joints and connections	2 - 4
Traps	3 - 4
Cleanouts	3 - 4

Interceptors	4 - 8
Fixtures	8 - 12
Hangers and supports	2 - 6
Indirect wastes	4 - 8
Water supply and distribution	8 - 12
Drainage systems	8 - 12
Vents and venting	8 - 12
Storm drains	1 - 3

Part II (open book, 30 questions, one hour, 20% of grade)

Subject	Number of questions
Fitting identification	8 - 12
General knowledge	6 - 10
Developed length	5 - 10
Drainage calculations	1 - 3
Vent calculations	2 - 4

Part III (open book, 40 questions, three hours, 40% of grade)

Subject	Number of questions
Isometric analysis	40

If you have a Plumbing license in North Carolina, you may be able to get your South Carolina license without taking the exam. Check with the Board on this.

Heating

The Heating exam is open book, three hours long, with 60 questions:

Subject	Number of questions
Code compliance	7 - 15
HARV general	3 - 7
HARV controls	5 - 10
HARV load calculation	8 - 12
Fuel gas systems	5 - 10
Piping	10 - 20
Ducts	5 - 10
Boilers — low pressure	10 - 20

If you have an Air Conditioning license in Georgia, a Class I Heating license in North Carolina or a Class A Air Conditioning license in Texas, you may be able to get your South Carolina license without taking the exam. Check with the Board on this.

Packaged Equipment

The Packaged Equipment exam has two parts:

Part I (open book, 50 questions, three hours, 50% of grade)

Subject	Number of questions
Code compliance	8 - 15
HARV general	10 - 15
Fuel gas systems	5 - 10
HARV load calculation	8 - 12
Insulation	4 - 6
Safety	3 - 6

Part II (open book, 50 questions, three hours, 50% of grade)

Subject	Number of questions
HARV maintenance	8 - 15
HARV controls	5 - 10
Safety — refrigeration	5 - 10
Ducts	5 - 10
Piping	8 - 15
Sheet metal	2 - 5
Psychrometric analysis	3 - 5

If you have a Class III Heating license in North Carolina or a Class B Air Conditioning license in Texas, you may be able to get your South Carolina license without taking the exam. Check with the Board on this.

Air Conditioning

The Air Conditioning exam has two parts:

Part I (open book, 60 questions, three hours, 50% of grade)

Subject	Number of questions
Code compliance	10 - 15
HARV general	10 - 20
Fuel gas systems	5 - 10
Safety	3 - 7
HARV load calculation	8 - 12

Part II (open book, 60 questions, three hours, 50% of grade)

Subject	Number of questions
HARV maintenance	10 - 15
HARV controls	8 - 15
Piping	9 - 18
Ducts	8 - 12
Psychrometric analysis	4 - 8
Insulation	4 - 8
Safety — refrigeration	4 - 8

If you have an Air Conditioning license in Georgia, a Class II Heating license in North Carolina or a Class B Air Conditioning license in Texas, you may be able to get your South Carolina license without taking the exam. Check with the Board on this.

Electrical

The Electrical exam has three parts:

Part I (closed book, one hour, 50 questions) — 21% of total grade

Subject	Number of questions
General theory	24 - 26
NEC chapter 1	6 - 8
NEC chapter 2	1 - 3
NEC chapter 3	0 - 2
NEC chapter 90	0 - 2
NEC chapter 4	6 - 8
NEC chapter 5	2 - 4
NEC chapter 6	2 - 4

Part II (open book, two hours, 70 questions) — 29% of total grade

Subject	Number of questions
General theory	8 - 10
NEC chapter 1	1 - 3
NEC chapter 2	9 - 11
NEC chapter 3	12 - 14
NEC chapter 4	16 - 18
NEC chapter 5	4 - 6
NEC chapter 6	7 - 9
NEC chapter 7	0 - 2
NEC chapter 8	1 - 3
NEC chapter 9	2 - 4

Part III (open book, three hours, 30 questions) — 50% of total grade

Subject	Number of questions
Service	4 - 6
Voltage drop	2 - 4
3Ø 208V delta and wye loads	2 - 4
Conduit fill	3 - 5
Conductor ampacity	2 - 4
Motors (branch circuit)	0 - 2
Motors (feeders)	0 - 2
Motors (protection)	0 - 2
Motors (control)	2 - 4
Appliance loads	1 - 3
Transformers, 3Ø transformer calculation	1 - 3
Grounding conductor and equipment ground	0 - 2
Power factor, power, and volt-amps	0 - 2

If you have an electrician's license in Alabama, Georgia, or North Carolina, you may be able to get your South Carolina license without taking the exam. Check with the Board on this.

Lightning Protection Systems

The Lightning Protection Systems exam is open book, three hours long, with 50 questions:

Subject	Number of questions
NFPA #780	27 - 29
General theory	10 - 12
NEC	8 - 14

Recommended Reading for the General or Mechanical Contractor's Exam

Building Contractor's Exam

📖 *Standard Building Code 1994*, Southern Building Code Congress International, Inc., 900 Montclair Road, Birmingham, AL 35213-1206

📖 *Code of Federal Regulations - Title 29, Part 1926 (OSHA),* July 1995, Superintendent of Documents, U.S. Government Printing Office, Washington, DC 20402

Recommended Reading for the General or Mechanical Contractor's Exam (continued)

Building Contractor's Exam (continued)

📖 *Design and Control of Concrete Mixtures,* 1988/1990, 13th edition, Portland Cement Association, 5420 Old Orchard Road, Skokie, IL 60077-1083

📖 *Concrete Masonry Handbook,* 1991, 5th edition, Portland Cement Association, 5420 Old Orchard Road, Skokie, IL 60077-1083

📖 *Handling and Erection of Steel Joists and Joist Girders, Technical Digest No. 9,* 1987, Steel Joist Institute, 1205 48th Avenue, North - Suite A, Myrtle Beach, SC 29577

📖 *Code of Standard Practice for Steel Buildings and Bridges,* 1992, American Institute of Steel Construction, P. O. Box 806276, Publication Department, Chicago, IL 60680-4124

📖 *Placing Reinforcing Bars, Recommended Practices,* 1992, 6th edition, Concrete Reinforcing Steel Institute, P. O. Box 6996, Alpharetta, GA 30239-6996

📖 *Span Tables for Joists and Rafters, 1993 with 1992 supplements,* American Forest and Paper Association, 1111 19th Street NW, Washington, DC 20002

📖 *Recommended Specifications for the Application and Finishing of Gypsum Board-GA-216-93,* 1993, Gypsum Association, 810 First Street NE, Washington, DC 20002

📖 *ASTM C-94 Standard Specification for Ready Mixed Concrete,* 1994, ASTM, 100 Bar Harbour Drive, West Conshohocken, PA 19428-2959

General Contractor's Exam

📖 *Standard Building Code 1994,* Southern Building Code Congress International, Inc., 900 Montclair Road, Birmingham, AL 35213-1206

📖 *Code of Federal Regulations - Title 29, Part 1926 (OSHA),* July 1995, Superintendent of Documents, U.S. Government Printing Office, Washington, DC 20402

📖 *Design and Control of Concrete Mixtures,* 1988/1990, 13th edition, Portland Cement Association, 5420 Old Orchard Road, Skokie, IL 60077-1083

📖 *Concrete Masonry Handbook,* 1991, 5th edition, Portland Cement Association, 5420 Old Orchard Road, Skokie, IL 60077-1083

📖 *Handling and Erection of Steel Joists and Joist Girders, Technical Digest No. 9,* 1987, Steel Joist Institute, 1205 48th Avenue, North - Suite A, Myrtle Beach, SC 29577

Recommended Reading for the General or Mechanical Contractor's Exam (continued)

General Contractor's Exam (continued)

- *Code of Standard Practice for Steel Buildings and Bridges,* 1992, American Institute of Steel Construction, P. O. Box 806276, Publication Department, Chicago, IL 60680-4124

- *Specification for Structural Joints Using ASTM A325 or A490 Bolts, ASD,* 1986, American Institute of Steel Construction, P. O. Box 806276, Publication Department, Chicago, IL 60680-4124

- *Placing Reinforcing Bars, Recommended Practices,* 1992, 6th edition, Concrete Reinforcing Steel Institute, P. O. Box 6996, Alpharetta, GA 30239-6996

- *Span Tables for Joists and Rafters,* 1993 with 1992 supplements, American Forest and Paper Association, 1111 19th Street NW, Washington, DC 20002

- *Recommended Specifications for the Application and Finishing of Gypsum Board-GA-216-93,* 1993, Gypsum Association, 810 First Street NE, Washington, DC 20002

- *ASTM C-94 Standard Specification for Ready Mixed Concrete,* 1994, ASTM, 100 Bar Harbour Drive, West Conshohocken, PA 19428-2959

- *Formwork for Concrete,* 1995, Revised, American Concrete Institute, P. O. Box 19150, Detroit, MI 48219-0150

- *Commentary and Recommendations for Handling, Installing and Bracing Metal Plate Connected Wood Trusses,* HIB-91, 1991, Truss Plate Institute, 583 D'Onofrio Drive, Madison, WI 53719

Highway Grading Exam

- *Code of Federal Regulations - Title 29, Part 1926 (OSHA),* July 1995, Superintendent of Documents, U.S. Government Printing Office, Washington, DC 20402

- *Excavation & Grading Handbook,* 1991, Nicholas E. Capachi, Craftsman Book Company, 6058 Corte del Cedro, P.O. Box 6500, Carlsbad, CA 92018

- *Standard Specifications for Highway Construction,* South Carolina State Highway Department, 1986, SC Department of Highways & Public Transportation, P. O. Box 191, Columbia, SC 29202

- *Construction Planning, Equipment and Methods,* 1995, 5th edition, R. L. Peurifoy, McGraw-Hill Publishing, Blue Ridge Summit, PA 17294

- *Formwork for Concrete,* 1995, Revised, American Concrete Institute, P. O. Box 19150, Detroit, MI 48219-0150

Recommended Reading for the General or Mechanical Contractor's Exam (continued)

Highway Grading Exam (continued)

📖 *Commentary and Recommendations for Handling, Installing and Bracing Metal Plate Connected Wood Trusses, HIB-91,* 1991, Truss Plate Institute, 583 D'Onofrio Drive, Madison, WI 53719

Highway Bridges Exam

📖 *Code of Federal Regulations - Title 29, Part 1926 (OSHA),* July 1995, Superintendent of Documents, U.S. Government Printing Office, Washington, DC 20402

📖 *Design and Control of Concrete Mixtures, 1988/1990,* 13th edition, Portland Cement Association, 5420 Old Orchard Road, Skokie, IL 60077-1083

📖 *ASTM C-94 Standard Specification for Ready Mixed Concrete,* 1994, ASTM, 100 Bar Harbour Drive, West Conshohocken, PA 19428-2959

📖 *Placing Reinforcing Bars, Recommended Practices,* 1992, 6th edition, Concrete Reinforcing Steel Institute, P. O. Box 6996, Alpharetta, GA 30239-6996

📖 *Formwork for Concrete,* 1995, Revised, American Concrete Institute, P. O. Box 19150, Detroit, MI 48219-0150

📖 *Hot Weather Concreting,* ACI 305R-91, 1991, American Concrete Institute, P. O. Box 19150, Detroit, MI 48219-0150

📖 *Cold Weather Concreting,* ACI 306R-88, 1988, American Concrete Institute, P. O. Box 19150, Detroit, MI 48219-0150

📖 *Manual of Steel Construction,* 1989, 9th edition, American Institute of Steel Construction, P. O. Box 806276, Chicago, IL 60680-4124

📖 *Handbook of Rigging,* 1988 4th edition, W. E. Rossnagel, McGraw-Hill Publishing, Inc., Blue Ridge Summit, PA 17294

📖 *MS-5 - Introduction to Asphalt,* 1986, 8th edition, Asphalt Institute, P. O. Box 14052, Lexington, KY 40512-4052

Highway Paving Exam

📖 *Construction Planning, Equipment and Methods,* 1995, 5th edition, R. L. Peurifoy, McGraw-Hill Publishing, Blue Ridge Summit, PA 17294

📖 *MS-5 - Introduction to Asphalt,* 1986, 8th edition, Asphalt Institute, P. O. Box 14052, Lexington, KY 40512-4052

Recommended Reading for the General or Mechanical Contractor's Exam (continued)

Highway Paving Exam (continued)

📖 *MS-16 - Asphalt in Pavement Maintenance,* 1987, 2nd edition, Asphalt Institute, P. O. Box 14052, Lexington, KY 40512-4052

📖 *MS-19 - A Basic Asphalt Emulsion Manual,* 1987, 2nd edition, Asphalt Institute, P. O. Box 14052, Lexington, KY 40512-4052

📖 *Design and Control of Concrete Mixtures,* 1988/1990, 13th edition, Portland Cement Association, 5420 Old Orchard Road, Skokie, IL 60077-1083

📖 *Cement Mason's Guide,* 1990, Portland Cement Association, 5420 Old Orchard Road, Skokie, IL 60077-1083

Public Utilities Gas Lines Exam

📖 *Standard Gas Code 1994,* Southern Building Code Congress International, Inc., 900 Montclair Road, Birmingham, AL 35213-1206

📖 *Standard Mechanical Code 1994* , Southern Building Code Congress International, Inc., 900 Montclair Road, Birmingham, AL 35213-1206

📖 *Code of Federal Regulations - Title 29, Part 1926 (OSHA),* July 1995, Superintendent of Documents, U.S. Government Printing Office, Washington, DC 20402

📖 *NFPA 30 - Flammable and Combustible Liquids Code,* 1992, National Fire Protection Association, 1 Batterymarch Park, P. O. Box 9101, Quincy, MA 02269

📖 *ASME B31.2 - Fuel Gas Piping,* 1968, American Society of Mechanical Engineers, 22 Law Drive, P. O. Box 2300, Fairfield, NJ 07007-2300

📖 *ASME B31.8- Gas Transmission and Distribution Piping Systems,* 1992, American Society of Mechanical Engineers, 22 Law Drive, P. O. Box 2300, Fairfield, NJ 07007-2300

📖 *Pipefitters Handbook,* 1967, 3rd edition, Forest Lindsey, Industrial Press, Inc., 200 Madison Avenue, New York, NY 10016

📖 *Pipe Welding Procedures,* 1973, H. Rampaul, Industrial Press, Inc., 200 Madison Avenue, New York, NY 10016

Recommended Reading for the General or Mechanical Contractor's Exam (continued)

Public Utilities Water and Sewer Lines Exam

📖 *Code of Federal Regulations - Title 29,* Part 1926 (OSHA), July 1995, Superintendent of Documents, U.S. Government Printing Office, Washington, DC 20402

📖 *Clay Pipe Engineering Manual,* 1995, National Clay Pipe Institute, P. O. Box 759, Lake Geneva, WI 53147

📖 *Clay Pipe Installation Handbook,* 1994, National Clay Pipe Institute, P. O. Box 759, Lake Geneva, WI 53147

📖 *A Guide for the Installation of Ductile Iron Pipe,* 1988, Ductile Iron Pipe Research Association, 245 Riverchase Parkway, East, Suite O, Birmingham, AL 35244

📖 *AWWA Standard for Installation of Ductile Iron Water Mains,* AWWA C-600-93, 1993, American Water Works Association, 6666 W. Quincy Ave., Denver, CO 80235

📖 *Concrete Pipe Installation Manual,* 1995, American Concrete Pipe Association, 8300 Boone Blvd., Suite 400, Vienna, VA 22182

📖 *Excavation & Grading Handbook,* 1991, Nicholas E. Capachi, Craftsman Book Company, 6058 Corte del Cedro, P.O. Box 6500, Carlsbad, CA 92018

Public Utilities Water and Sewer Plants Exam

📖 *Code of Federal Regulations - Title 29, Part 1926 (OSHA),* July 1995, Superintendent of Documents, U.S. Government Printing Office, Washington, DC 20402

📖 *Design of Municipal Wastewater Treatment Plants,* 1992, Water Environment Federation, 601 Wythe St., Alexandria, VA 22314-1994

📖 *Design and Control of Concrete Mixtures,* 1988/1990, 13th edition, Portland Cement Association, 5420 Old Orchard Road, Skokie, IL 60077-1083

📖 *ASTM C-94 Standard Specification for Ready Mixed Concrete,* 1994, ASTM, 100 Bar Harbour Drive, West Conshohocken, PA 19428-2959

📖 *Placing Reinforcing Bars, Recommended Practices,* 1992, 6th edition, Concrete Reinforcing Steel Institute, P. O. Box 6996, Alpharetta, GA 30239-6996

📖 *Excavation & Grading Handbook,* 1991, Nicholas E. Capachi, Craftsman Book Company, 6058 Corte del Cedro, P.O. Box 6500, Carlsbad, CA 92018

Recommended Reading for the General or Mechanical Contractor's Exam (continued)

Public Utilities Electrical Exam

📖 *Cableman's and Lineman's Handbook,* Kurtz and Shoemaker, McGraw-Hill Company, Blue Ridge Summit, PA 17294

📖 *ANSI C2 - 1993 - National Electric Safety Code,* 1993, ANSI, 11 West 42nd St., New York, NY 10036

Refrigeration Contractor's Exam

📖 *Standard Mechanical Code 1994 ,* Southern Building Code Congress International, Inc., 900 Montclair Road, Birmingham, AL 35213-1206

📖 *Code of Federal Regulations - Title 29, Part 1926 (OSHA),* July 1995, Superintendent of Documents, U.S. Government Printing Office, Washington, DC 20402

📖 *ANSI/ASHRAE 15-94 - Safety Code for Mechanical Refrigeration,* 1994 ANSI, 11 West 42nd St., New York, NY 10036

📖 *Trane Air Conditioning Manual,* 1965, Trane Company, 8929 Western Way, Suite #1, Jacksonville, FL 32256

📖 *Refrigeration and Air Conditioning,* 1987, 2nd edition, ARI, Prentice Hall, P. O. Box 11071, Des Moines, IA 50336-1071

📖 *Pipefitters Handbook,* 1967, 3rd edition, Forest Lindsey, Industrial Press, Inc., 200 Madison Avenue, New York, NY 10016

📖 *System Design Manual - Part III,* Piping Design, Carrier, 1960, Carrier Air Conditioning Company, Carrier Parkway, P. O. Box 4808, Syracuse, NY 13221

Concrete Contractor's Exam

📖 *Standard Building Code 1994,* Southern Building Code Congress International, Inc., 900 Montclair Road, Birmingham, AL 35213-1206

📖 *Code of Federal Regulations - Title 29, Part 1926 (OSHA),* July 1995, Superintendent of Documents, U.S. Government Printing Office, Washington, DC 20402

📖 *Placing Reinforcing Bars, Recommended Practices,* 1992, 6th edition, Concrete Reinforcing Steel Institute, P. O. Box 6996, Alpharetta, GA 30239-6996

📖 *Design and Control of Concrete Mixtures,* 1988/1990, 13th edition, Portland Cement Association, 5420 Old Orchard Road, Skokie, IL 60077-1083

Recommended Reading for the General or Mechanical Contractor's Exam (continued)

Concrete Contractor's Exam (continued)

📖 *Cement Mason's Guide,* 1990, Portland Cement Association, 5420 Old Orchard Road, Skokie, IL 60077-1083

📖 *ASTM C-94 Standard Specification for Ready Mixed Concrete,* 1994, ASTM, 100 Bar Harbour Drive, West Conshohocken, PA 19428-2959

📖 *Formwork for Concrete,* 1995, Revised, American Concrete Institute, P. O. Box 19150, Detroit, MI 48219-0150

Marine Contractor's Exam

📖 *Code of Federal Regulations - Title 29, Part 1926 (OSHA),* July 1995, Superintendent of Documents, U.S. Government Printing Office, Washington, DC 20402

📖 *Construction Planning, Equipment and Methods,* 1995, 5th edition, R. L. Peurifoy, McGraw-Hill Publishing, Blue Ridge Summit, PA 17294

📖 *Design and Control of Concrete Mixtures,* 1988/1990, 13th edition, Portland Cement Association, 5420 Old Orchard Road, Skokie, IL 60077-1083

📖 *ASTM C-94 Standard Specification for Ready Mixed Concrete,* 1994, ASTM, 100 Bar Harbour Drive, West Conshohocken, PA 19428-2959

📖 *Placing Reinforcing Bars, Recommended Practices,* 1992, 6th edition, Concrete Reinforcing Steel Institute, P. O. Box 6996, Alpharetta, GA 30239-6996

Process Piping Contractor's Exam

📖 *Standard Gas Code 1994,* Southern Building Code Congress International, Inc., 900 Montclair Road, Birmingham, AL 35213-1206

📖 *Standard Mechanical Code 1994 ,* Southern Building Code Congress International, Inc., 900 Montclair Road, Birmingham, AL 35213-1206

📖 *Code of Federal Regulations - Title 29, Part 1926 (OSHA),* July 1995, Superintendent of Documents, U.S. Government Printing Office, Washington, DC 20402

📖 *ASME B31.8- Gas Transmission and Distribution Piping Systems,* 1992, American Society of Mechanical Engineers, 22 Law Drive, P. O. Box 2300, Fairfield, NJ 07007-2300

📖 *Pipefitters Handbook,* 1967, 3rd edition, Forest Lindsey, Industrial Press, Inc., 200 Madison Avenue, New York, NY 10016

Recommended Reading for the General or Mechanical Contractor's Exam (continued)

Process Piping Contractor's Exam (continued)

📖 *Pipe Welding Procedures,* 1973, H. Rampaul, Industrial Press, Inc., 200 Madison Avenue, New York, NY 10016

📖 *System Design Manual - Part III,* Piping Design, Carrier, 1960, Carrier Air Conditioning Company, Carrier Parkway, P. O. Box 4808, Syracuse, NY 13221

Roofing Contractor's Exam

📖 *Standard Building Code 1994,* Southern Building Code Congress International, Inc., 900 Montclair Road, Birmingham, AL 35213-1206

📖 *Code of Federal Regulations - Title 29, Part 1926 (OSHA),* July 1995, Superintendent of Documents, U.S. Government Printing Office, Washington, DC 20402

📖 *Roofing Construction & Estimating,* 1995, Daniel Atcheson, Craftsman Book Company, 6058 Corte del Cedro, P.O. Box 6500, Carlsbad, CA 92018

📖 *Residential Asphalt Roofing Manual,* 1993 revised, Asphalt Roofing Manufacturer's Association, 6000 Executive Blvd., Suite 201, Rockville, MD 20852

📖 *NRCA Roofing and Waterproofing Manual,* 1989, 3rd edition, National Roofing Contractor's Association, 10255 W. Higgins, Rd., Suite #600, Rosemont, IL 60018-5607

Swimming Pool Contractor's Exam

📖 *Regulation 61-51, Public Swimming Pools,* revised 1992, South Carolina Department of Health and Environmental Control, Bureau of Drinking Water, 2600 Bull St., Columbia, SC 29201

📖 *Code of Federal Regulations - Title 29, Part 1926 (OSHA),* July 1995, Superintendent of Documents, U.S. Government Printing Office, Washington, DC 20402

📖 *Article 680, National Electrical Code: Swimming Pool Wiring,* 1993, National Fire Protection Association, 1 Batterymarch Park, P. O. Box 9101, Quincy, MA 02269

📖 *Guide to Shotcrete, ACI 506R-90,* 1990, American Concrete Institute, P. O. Box 19150, Detroit, MI 48219-0150

📖 *Design and Control of Concrete Mixtures,* 1988/1990, 13th edition, Portland Cement Association, 5420 Old Orchard Road, Skokie, IL 60077-1083

Recommended Reading for the General or Mechanical Contractor's Exam (continued)

Swimming Pool Contractor's Exam (continued)

📖 *Placing Reinforcing Bars, Recommended Practices,* 1992, 6th edition, Concrete Reinforcing Steel Institute, P. O. Box 6996, Alpharetta, GA 30239-6996

📖 *Pool/Spa Operators Handbook,* 1990, National Swimming Pool Foundation, 10803 Gulfdale, Suite 300, San Antonio, TX 78216

Wood Frames Structures I Contractor's Exam

📖 *Standard Building Code 1994 ,* Southern Building Code Congress International, Inc., 900 Montclair Road, Birmingham, AL 35213-1206

📖 *Code of Federal Regulations - Title 29, Part 1926 (OSHA),* July 1995, Superintendent of Documents, U.S. Government Printing Office, Washington, DC 20402

📖 *Carpentry and Building Construction,* 1993, Feirer, Hutchings & Feirer, McGraw-Hill Inc., Box 543, Blacklick, OH 43004-0543

📖 *Commentary and Recommendations for Handling, Installing and Bracing Metal Plate Connected Wood Trusses, HIB-91,* 1991, Truss Plate Institute, 583 D'Onofrio Drive, Madison, WI 53719

Wood Frames Structures II Contractor's Exam

📖 *Standard Building Code 1994 ,* Southern Building Code Congress International, Inc., 900 Montclair Road, Birmingham, AL 35213-1206

📖 *Carpentry and Building Construction,* 1993, Feirer, Hutchings & Feirer, McGraw-Hill Inc., Box 543, Blacklick, OH 43004-0543

📖 *Commentary and Recommendations for Handling, Installing and Bracing Metal Plate Connected Wood Trusses,* HIB-91, 1991, Truss Plate Institute, 583 D'Onofrio Drive, Madison, WI 53719

Plumbing I Contractor's Exam

📖 *South Carolina Code of Laws, Chapter 11 (40-11-10 to 40-11-340) & Chapter 29 (29-1 to 29-40): An Act Regulating General & Mechanical Contracting,* 1993, SC Department of Labor, Licensing Regulation, Contractors Licensing Board, P. O. Box 11329, Columbia, SC 29211-1329

📖 *Standard Plumbing Code 1994,* Southern Building Code Congress International, Inc., 900 Montclair Road, Birmingham, AL 35213-1206

Recommended Reading for the General or Mechanical Contractor's Exam (continued)

Plumbing I Contractor's Exam (continued)

📖 *Mathematics for Plumbers and Pipefitters,* 1989, 5th edition, D'Arcangelo, D'Arcangelo and Guest, Delmar Publishers, P. O. Box 6904, Florence, KY 41022

📖 *Plumbing Technology: Design and Installation,* 1994, 2nd edition, Lee Smith, Delmar Publishers, P. O. Box 6904, Florence, KY 41022

📖 *Blueprint Reading for Plumbers - Residential and Commercial,* 1989, 5th edition, D'Arcangelo, D'Arcangelo and Guest, Delmar Publishers, P. O. Box 6904, Florence, KY 41022

📖 *Plumbing,* L. V. Ripka, American Technical Publishers, 1155 West 175th Street, Homewood, IL 60403

📖 *Workbook for Plumbing Installation and Design,* 2nd edition, Charles Hollar, American Technical Publishers, 1155 West 175th Street, Homewood, IL 60403

Heating Contractor's Exam

📖 *Fibrous Glass Duct Construction Standards,* 1993, 2nd edition, North American Insulation Manufacturers Assoc., 44 Canal Center Plaza, Suite #310, Alexandria, VA 22314

📖 *Standard Mechanical Code 1994,* Southern Building Code Congress International, Inc., 900 Montclair Road, Birmingham, AL 35213-1206

📖 *Standard Gas Code 1994,* Southern Building Code Congress International, Inc., 900 Montclair Road, Birmingham, AL 35213-1206

📖 *NFPA 54 - National Fuel Gas Code,* 1992, National Fire Protection Association, 1 Batterymarch Park, P. O. Box 9101, Quincy, MA 02269

📖 *Low Pressure Boilers,* 1994, F. M. Steingress, American Technical Publishers, 1155 West 175th Street, Homewood, IL 60403

📖 *System Design Manual - Part I, Load Estimating, Carrier,* 1960, Carrier Air Conditioning Company, Carrier Parkway, P. O. Box 4808, Syracuse, NY 13221

📖 *System Design Manual - Part II, Air Distribution, Carrier,* 1960, Carrier Air Conditioning Company, Carrier Parkway, P. O. Box 4808, Syracuse, NY 13221

📖 *System Design Manual - Part III, Piping Design, Carrier,* 1960, Carrier Air Conditioning Company, Carrier Parkway, P. O. Box 4808, Syracuse, NY 13221

Recommended Reading for the General or Mechanical Contractor's Exam (continued)

Heating Contractor's Exam (continued)

📖 *Refrigeration and Air Conditioning,* 1995, 3rd edition, ARI, Prentice Hall, P. O. Box 11071, Des Moines, IA 50336-1071

📖 *Trane Air Conditioning Manual,* 1965, Trane Company, 8929 Western Way, Suite #1, Jacksonville, FL 32256

📖 *Trane Ductulator,* 1976, Trane Company, 8929 Western Way, Suite #1, Jacksonville, FL 32256

Packaged Equipment Contractor's Exam

📖 *South Carolina Code of Laws, Chapter 11 (40-11-10 to 40-11-340) & Chapter 29 (29-1 to 29-40): An Act Regulating General & Mechanical Contracting,* 1993, SC Department of Labor, Licensing Regulation, Contractors Licensing Board, P. O. Box 11329, Columbia, SC 29211-1329

📖 *Standard Mechanical Code 1994,* Southern Building Code Congress International, Inc., 900 Montclair Road, Birmingham, AL 35213-1206

📖 *Code of Federal Regulations - Title 29, Part 1926 (OSHA),* July 1995, Superintendent of Documents, U.S. Government Printing Office, Washington, DC 20402

📖 *NFPA 54 - National Fuel Gas Code,* 1992, National Fire Protection Association, 1 Batterymarch Park, P. O. Box 9101, Quincy, MA 02269

📖 *ANSI/ASHRAE 15-94 - Safety Code for Mechanical Refrigeration,* 1994 ANSI, 11 West 42nd St., New York, NY 10036

📖 *Refrigeration and Air Conditioning,* 1987, 2nd edition, ARI, Prentice Hall, P. O. Box 11071, Des Moines, IA 50336-1071

📖 *Trane Air Conditioning Manual,* 1965, Trane Company, 8929 Western Way, Suite #1, Jacksonville, FL 32256

📖 *Trane Ductulator,* 1976, Trane Company, 8929 Western Way, Suite #1, Jacksonville, FL 32256

📖 *Manual J - Load Calculation for Residential Winter and Summer Air Conditioning,* 1986, 7th edition, Air Conditioning Contractors of America, 1712 New Hampshire Ave., NW, Washington, DC 20009

📖 *Pipefitters Handbook,* 1967, 3rd edition, Forest Lindsey, Industrial Press, Inc., 200 Madison Avenue, New York, NY 10016

📖 *System Design Manual - Part III, Piping Design,* Carrier, 1960, Carrier Air Conditioning Company, Carrier Parkway, P. O. Box 4808, Syracuse, NY 13221

Recommended Reading for the General or Mechanical Contractor's Exam (continued)

Air Conditioning Contractor's Exam

📖 *South Carolina Code of Laws, Chapter 11 (40-11-10 to 40-11-340) & Chapter 29 (29-1 to 29-40): An Act Regulating General & Mechanical Contracting,* 1993, SC Department of Labor, Licensing Regulation, Contractors Licensing Board, P. O. Box 11329, Columbia, SC 29211-1329

📖 *Standard Mechanical Code 1994,* Southern Building Code Congress International, Inc., 900 Montclair Road, Birmingham, AL 35213-1206

📖 *Standard Gas Code 1994,* Southern Building Code Congress International, Inc., 900 Montclair Road, Birmingham, AL 35213-1206

📖 *Code of Federal Regulations - Title 29, Part 1926 (OSHA),* July 1995, Superintendent of Documents, U.S. Government Printing Office, Washington, DC 20402

📖 *NFPA 91- Exhaust Systems for Air Conveying of Materials,* 1995, National Fire Protection Association, 1 Batterymarch Park, Box 9101, Quincy, MA 02269-9101

📖 *NFPA 96 - Standard for Ventilation Control and Fire Protection of Commercial Cooking Operations,* 1994, National Fire Protection Association, 1 Batterymarch Park, Box 9101, Quincy, MA 02269-9101

📖 *ANSI/ASHRAE 15-94 - Safety Code for Mechanical Refrigeration,* 1994 ANSI, 11 West 42nd St., New York, NY 10036

📖 *Refrigeration and Air Conditioning,* 1987, 2nd edition, ARI, Prentice Hall, P. O. Box 11071, Des Moines, IA 50336-1071

📖 *Manual N - Load Calculation for Commercial Summer and Winter Air Conditioning,* 1988, 4th edition, Air Conditioning Contractors of America, 1712 New Hampshire Ave., NW, Washington, DC 20009

📖 *Trane Air Conditioning Manual,* 1965, Trane Company, 8929 Western Way, Suite #1, Jacksonville, FL 32256

📖 *Trane Ductulator,* 1976, Trane Company, 8929 Western Way, Suite #1, Jacksonville, FL 32256

📖 *Pipefitters Handbook,* 1967, 3rd edition, Forest Lindsey, Industrial Press, Inc., 200 Madison Avenue, New York, NY 10016

📖 *System Design Manual - Part I, Load Estimating,* Carrier, 1960, Carrier Air Conditioning Company, Carrier Parkway, P. O. Box 4808, Syracuse, NY 13221

📖 *System Design Manual - Part III, Piping Design,* Carrier, 1960, Carrier Air Conditioning Company, Carrier Parkway, P. O. Box 4808, Syracuse, NY 13221

Recommended Reading for the General or Mechanical Contractor's Exam (continued)

Electrical Contractor's Exam

📖 *NFPA 70 - National Electrical Code,* 1996 edition, National Fire Protection Association, 1 Batterymarch Park, P. O. Box 9101, Quincy, MA 02269

📖 *American Electricians Handbook,* 1996, 13th edition, Croft/Summers, McGraw-Hill Inc., Box 543, Blacklick, OH 43004-0543

📖 *National Electrical Code Blueprint Reading,* K. L. Gebert, American Technical Publishers, 1155 West 175th Street, Homewood, IL 60403

📖 *Electrical Review for Electricians,* 9th edition, 1996, 9th edition, J. Morris Trimmer and Charles Pardue, Construction Bookstore, 100 Enterprise Place, Dover, DE 19903-7029

Lightning Protection Systems Exam

📖 *NFPA 70 - National Electrical Code,* 1996 edition, National Fire Protection Association, 1 Batterymarch Park, P. O. Box 9101, Quincy, MA 02269

📖 *NFPA 780 - Lighting Protection Code,* 1992, National Fire Protection Association, 1 Batterymarch Park, P. O. Box 9101, Quincy, MA 02269

📖 *American Electricians Handbook,* 1996, 13th edition, Croft/Summers, McGraw-Hill Inc., Box 543, Blacklick, OH 43004-0543

Asbestos

To work on any asbestos or asbestos abatement project in South Carolina you must be licensed by the Department of Health and Environmental Control (DHEC). To get an application for a license contact:

Contact

Bureau of Air Quality
South Carolina DHEC
2600 Bull Street
Columbia, SC 29201
(803) 734-4517

Here's a table of the licenses DHEC issues and the cost of each:

Fee

License	Fee
Contractor	$100
Supervisor	$50

Worker	$10
Air Sampler	$100
Project designer	$100
Building inspector	$100
Management planner	$100

All licenses except the contractor license require you to successfully complete Department-approved asbestos training courses. To get a contractor license you must certify that you employ a licensed supervisor. If you have a license and/or training from another state, the Department may accept it. Check with them about this. All licenses are good for one year.

Department of Transportation (DOT)

To bid on any South Carolina Department of Transportation construction project, you must be prequalified by the Department. You must also have a General Contractor and Bidder license from the South Carolina Contractor's Licensing Board to win a contract with the Department. To get the Prime Contractor Prequalification Questionnaire, contact:

Contact

Contract Administrator
South Carolina Department of Transportation
955 Park Street
P. O. Box 191
Columbia, SC 29201-0191
(803) 737-1249

The Transportation Department will ask you for these items:

- financial statement
- which types of work you want to be prequalified for
- disadvantaged business status of your firm
- equal employment opportunity affidavit
- your current and past work projects as a prime contractor
- what equipment your firm owns or leases

You'll also be asked which of the following types of work you want to be prequalified in:

Paving	Hydraulic embankments
Grading	Jetties or groins
Bituminous surfacing	Electrical work
Bridges	Signs
Seeding and grassing	General

The Department will use this information to give your company a prequalification rating based on your verified experience, net liquid assets, responsibility record, and available equipment for the types of work you select.

Out-of-State Corporations

Corporations doing business in South Carolina should register with the South Carolina Secretary of State to do business in the state. For information contact:

Contact

Corporation Department
Office of the Secretary of State
P. O. Box 11350
Columbia, SC 29211
(803) 734-2170
Fax (803) 734-2164

South Dakota

South Dakota only certifies or licenses asbestos abatement, electrical and plumbing contractors.

Asbestos Abatement Certificates

Unless you work only on small scale, short duration projects, you need to be certified to do asbestos work in South Dakota. To get an application, contact:

Contact

South Dakota Department of Environment and Natural Resources
Attention: Asbestos Coordinator
Office of Waste Management
523 East Capitol Avenue
Pierre, SD 57501-3181
(605) 773-3153

The Department issues the following types of certificates:

Inspector	Abatement contractor
Management planner	Abatement supervisor
Abatement project designer	Abatement worker

When you apply for a certificate, you must submit a signed statement that you have read, understood, and will comply with all state and federal rules on asbestos. You also must complete Department-approved training which includes passing a closed-book exam. Here's information on each type of certificate, the length of the Department-approved training course you need to get the certificate, and the number of multiple choice questions on the exam for the certificate:

Certificate	Length of training course	Number of questions
Inspector	Three days	50
Management planner	Three-day inspector training course plus two-day management planner training course	50
Project designer	Five days	100

Certificate	Length of training course	Number of questions
Contractor	Five days	100
Supervisor	Five days	100
Worker	Four days	50

If you can document that you've successfully completed an EPA-approved training course in another state, the Department will accept that training.

Asbestos abatement certificate fees: It will cost you $100 to get a certificate and it's good for one year.

Electrician's Licenses

To do electrical work in South Dakota, you need a license. To get an application, contact:

South Dakota Electrical Commission
302 S. Pawnee
c/o 500 E. Capitol Ave.
Pierre, SD 57501-5070
(605) 773-3573
(800) 233-7765
Fax: (605) 773-6213

The Commission issues electrical contractor, journeyman, and Class B licenses. The Class B license is limited to residential and farmstead electrical work. To get a contractor or Class B license, you must have public liability insurance for at least $300,000. You also need to post a $10,000 bond.

You must pass an exam to get any license. To qualify for the journeyman exam you need four years of experience as an apprentice, wiring, installing and repairing electrical apparatus and equipment. To qualify for the contractor exam you need 4,000 hours as a journeyman electrician. At least 2,000 hours of this time must be in commercial/industrial electrical work. To qualify for the Class B license you need two years of experience as a journeyman. The Commission gives all exams. They're based on the *National Electrical Code* and last five to six hours.

Electrician's license fees: It will cost you $40 to take a licensing exam. The contractor's license costs $100 and the journeyman or Class B license is $40. All fees are nonrefundable. All licenses are good for two years, expiring on June 30 each even-numbered year.

If you have a valid license in another state, the Commission may grant you a South Dakota license. You must have passed an exam to get your license and you must have work experience basically equal to South Dakota's. You must have held your license for at least one year. Currently the Commission accepts valid electrician licenses from these states:

Alaska (journeyman only)	Minnesota
North Dakota	Washington (journeyman only)
Idaho (journeyman only)	Nebraska
Utah (journeyman only)	Wyoming

You still must pay the application and license fees to get your license.

Plumber's Licenses

To do plumbing work in South Dakota, you need a license. To get an application, contact:

South Dakota State Plumbing Commission
P. O. Box 807
Pierre, SD 57501-0807
(605) 773-3429
Fax: (605) 773-5405

The Commission issues contractor and journeyman licenses. You need two years of experience as a journeyman to qualify for a contractor license and four years of experience as an apprentice to qualify for a journeyman license.

You also have to pass an exam for either license. The Commission gives the exam, which is based on the *Standard Plumbing Code*. The exam has three parts with multiple choice and essay questions and a drawing. It lasts about four hours.

Plumber's license fees: It will cost you $215 for a contractor license and $110 for a journeyman license. The license fee includes the cost of the exam. All licenses expire on December 31 each year.

The Commission has reciprocity agreements with Minnesota, North Dakota, Montana, and Colorado.

Department of Transportation (DOT)

To bid on South Dakota Department of Transportation projects you have to be prequalified by the Department. To get an application for prequalification, contact:

South Dakota Department of Transportation
700 East Broadway Avenue
Pierre, SD 57501-2586
(605) 773-3284
Fax: (605) 773-3921

Some of the details the Department will ask you about are your organization, personnel, financial condition, equipment, and experience. You'll be asked what type of work you want to be qualified for. Here are the types of work the Department uses:

Grading

Portland cement concrete paving

Asphalt concrete paving

Asphalt surface treatment

Gravel or crushed rock surfacing and base course

Crushing, stockpiling, salvage in-place asphalt surfacing

Bridges

Box culverts, precast multi-beam deck bridges

Lighting, signals

Signing, delineation

Rest area buildings

Erosion control

Incidental construction

Miscellaneous concrete construction

Bridge painting

Bituminous recycling, milling, salvaging

All types listed above

Prequalification is good for 18 months.

Out-of-State Corporations

If your company isn't incorporated under the laws of South Dakota, you have to get a Certificate of Authority from the Secretary of State to do business in the state. Contact the Secretary of State at:

Contact

Corporations
Office of Secretary of State
500 East Capitol
Pierre, SD 57501
(605) 773-4845
Fax: (605) 773-4550

Tennessee

You must have a license to do construction work in Tennessee. To get an application, contact:

Board for Licensing Contractors
500 James Robertson Parkway, Suite 110
Nashville, TN 37243-1150
(615) 741-8307
Fax: (615) 532-2868

You'll have to fill out an extensive application for the Board. Some important items you have to include are:

- one reference letter from a past client or employer who can comment on your construction work and experience or a code official who has inspected your work
- a current financial statement
- a copy of your charter if you're a Tennessee corporation or a Certificate of Authority from the Tennessee Secretary of State if you are incorporated outside Tennessee

You also have to complete a worksheet to determine monetary limits for your company. The limits are based on ten times your working capital or net worth. There are some possibilities for you to increase the limits with guaranty agreements and lines of credit.

You'll have to pass the Tennessee business and law exam, and possibly a trade exam, to get a license from the Board. The exams are given by Block & Associates. You can contact them at:

Block & Associates
2100 NW 53rd Avenue
Gainesville, FL 32653-2149
(800) 280-3926

The business and law exam has 50 multiple choice questions on state and federal laws and business, financial, and accounting practices. It lasts two hours. Here's a list of the subjects on the business and law exam and the approximate percentage of each subject on the exam:

Subject	Approximate percent of exam
Project management	20
Contract management	20
License law and rules	10
Financial management	10
Safety requirements	10
Employment laws	8
Payroll taxes	6
Risk management	6
Mechanics' lien law	6
Business organization	4

Block & Associates has prepared the *Tennessee Contractors Reference Manual* that you can use at the exam. They sell it for $35 plus tax and shipping. They also sell a practice business and law exam for $25 plus tax.

The Board has a lengthy list of the trades it licenses. For many of the trades, you'll also have to pass a trade exam. The trades which involve environmental work (asbestos abatement, underground storage tank installation, lead abatement, hazardous waste removal, and air, water, or soil remediation) all require you to have a current EPA training certificate.

In addition to a license from the Board, you must also register with the Department of Commerce and Insurance if you do any electrical work which requires inspection. To get an application, you can contact them at:

Contact

Department of Commerce and Insurance
Permits and Licenses Section
500 James Robertson Parkway, 3rd floor
Nashville, TN 37243-1159
(615) 741-1322
Fax: (615) 751-1583

Fee

Contractor's license fees: This registration costs $25 per year. After you pass the required exams, you have to be interviewed by the Board. This interview will last about 30 minutes. The Board will ask you about your company and your contracting experience.

It will cost you $150 nonrefundable to file your application and get your license if you pass the exams. Exams cost $45 each. A license is good for one year.

Here's information on all of the trades which require an exam, followed by a reference list for each exam. You can get the references listed from:

Contact

Professional Booksellers
2200 21st Avenue South, Suite 105
Nashville, TN 37212
(800) 572-8878

Commercial Building Contractor

The test is 60 multiple choice questions, four hours:

Subject	Approximate percent of exam
Carpentry	23
Steel and rebar	22
Concrete	17
Site work	12
Roofing	10
Masonry	8
Associated trades	8

Commercial/Industrial Contractor

The test is 80 multiple choice questions, four hours:

Subject	Approximate percent of exam
Structural steel	30
Concrete rebar	19
Site work	12
Carpentry	10
Masonry	10
Associated trades	8
Roofing	7
Industrial equipment, piping	4

Commercial/Industrial/Residential Contractor

The test is 80 multiple choice questions, four hours:

Subject	Approximate percent of exam
Concrete and rebar	20
Structural steel	19
Carpentry	18
Site work	18
Associated trades	10
Masonry	8
Roofing	7

Electrical Contractor

The test is 150 multiple choice questions, six hours:

Subject	Approximate percent of exam
Calculations	30
General use equipment	24
Electrical theory	14
Wiring methods, materials	11
Special equipment	5
Wiring and protection	5
Special occupancies	4
General code knowledge	4
Low voltage systems	3

High Voltage Electrical Contractor

The test is 20 multiple choice questions, one hour:

Subject	Approximate percent of exam
Distribution transformers	15
Overhead lines	15
Safety	15
Conductors	10
Electric supply stations	10
Grounding	10
Underground lines	10
Service drops, laterals	5
Poles and structures	5
Utilization equipment	5

HVAC Refrigeration Contractor

The test is 60 multiple choice questions, three hours:

Subject	Approximate percent of exam
Air conditioning	15
Warm air heating	10
Refrigeration	10
Steam, hot water boilers	10
Steamed, hot, chilled, condenser water piping	10
Ventilation	10
Ducts, flues, vents	10
Gas, fuel oil piping	10
Sizing and estimating	5
Testing, inspecting, balancing	5
Controls	5

Industrial Building Contractor

The test is 60 multiple choice questions, three hours:

Subject	Approximate percent of exam
Structural steel and rebar	35
Concrete	18
Carpentry	10
Excavation and site work	10
Roofing	10
Associated trades	10
Welding	4
Industrial equipment, piping	3

Residential/Small Commercial Contractor

The test is 60 multiple choice questions, three hours:

Subject	Approximate percent of exam
Carpentry	35
Concrete and rebar	15
Roofing	10
Masonry	10
Associated trades	10
Steel	10
Site work	10

Mechanical Contractor

Part I, 50% of total grade (60 multiple choice questions, three hours):

Subject	Approximate percent of exam
Code compliance	13
Fuel gas systems	17
Pollution control	5
HARV load calculation	15
Boilers - low pressure	8
Boilers - high pressure	8
Piping steam	17
Welding	10
Piping, power piping	7

Part II, 50% of total grade (60 multiple choice questions, three hours):

Subject	Approximate percent of exam
Fire sprinkler	8
Plumbing - general knowledge	12
Plumbing - basic principles	12
Plumbing - troubleshooting	10
General HARV	10
HARV maintenance	12
HARV controls	10
Safety	5
Piping	12
Insulation (piping only)	8
Psychrometric analysis	1

Residential Contractor

The test is 60 multiple choice questions, three hours:

Subject	Approximate percent of exam
Rough carpentry	20
Finish carpentry	18
Concrete and rebar	30
Excavation and site work	13
Roofing	12
Associated trades	12
Masonry	10

Plumbing Contractor

The test is 100 multiple choice questions, three hours:

Subject	Approximate percent of exam
Plumbing administration	6
Basic principles	3
Definition	4
General regulations	8
Materials	3
Joints and connections	3
Traps	3
Cleanouts	6
Interceptors	6
Fixtures	7

Subject	Approximate percent of exam
Hangers and supports	3
Water supply, distribution	7
Drainage systems	7
Storm drains	1
Fuel gas systems	5
Septic tanks	2
Drainage calculations	3
Vent calculations	3
Isometric analysis	12

Small Commercial Contractor

The test is 60 multiple choice questions, three hours:

Subject	Approximate percent of exam
Carpentry	35
Concrete and rebar	15
Site work	10
Roofing	10
Masonry	10
Steel	10
Associated trades	10

Recommended Reading for the Contractor's Exams

Commercial Building Contractor's Exam

📖 *SBCCI Standard Building Code 1994,* Southern Building Code Congress International, Inc., 900 Montclair Road, Birmingham, AL 35213-1206

📖 *NFPA 70 - National Electrical Code,* 1996 edition, National Fire Protection Association, 1 Batterymarch Park, P. O. Box 9101, Quincy, MA 02269

📖 *Code of Federal Regulations - Title 29, Part 1926 (OSHA),* July 1995, Superintendent of Documents, U.S. Government Printing Office, Washington, DC 20402

📖 *Carpentry and Building Construction,* 1993, Feirer, Hutchings & Feirer, McGraw-Hill Inc., Box 543, Blacklick, OH 43004-0543

📖 *Construction Materials and Processes,* 1990, Watson, McGraw-Hill Inc., Box 543, Blacklick, OH 43004-0543

📖 *Design and Control of Concrete Mixtures,* 1988/1990, 13th edition, Portland Cement Association, 5420 Old Orchard Road, Skokie, IL 60077-1083

Recommended Reading for the Contractor's Exams (continued)

Commercial Building Contractor's Exam (continued)

NRCA Roofing and Waterproofing Manual, 1996, 4th edition, National Roofing Contractor's Association, 10255 W. Higgins, Rd., Suite #600, Rosemont, IL 60018-5607

Principles and Practices of Heavy Construction, 1993, R. C. Smith & C.K. Andres, Prentice-Hall Co., P. O. Box 11071, Des Moines, IA 50336-1071

Excavation & Grading Handbook, 1991, Nicholas E. Capachi, Craftsman Book Company, 6058 Corte del Cedro, P.O. Box 6500, Carlsbad, CA 92018

Concrete Construction & Estimating, 1991, Craig Avery, Craftsman Book Company, 6058 Corte del Cedro, P.O. Box 6500, Carlsbad, CA 92018

Manual of Steel Construction, 1989, 9th edition, American Institute of Steel Construction, P. O. Box 806276, Chicago, IL 60680-4124

Handbook of Rigging, 1988 4th edition, W. E. Rossnagel, McGraw-Hill Publishing, Inc., Blue Ridge Summit, PA 17294

Construction of Structural Steel Building Frames , W. G. Rapp, Roger E. Krieger Publishing Co., P. O. Box 9542, Melbourne, FL 32952

Commercial/Industrial Contractor's Exam

SBCCI Standard Building Code 1994, Southern Building Code Congress International, Inc., 900 Montclair Road, Birmingham, AL 35213-1206

NFPA 70 - National Electrical Code, 1996 edition, National Fire Protection Association, 1 Batterymarch Park, P. O. Box 9101, Quincy, MA 02269

Code of Federal Regulations - Title 29, Part 1926 (OSHA), July 1995, Superintendent of Documents, U.S. Government Printing Office, Washington, DC 20402

Carpentry and Building Construction, 1993, Feirer, Hutchings & Feirer, McGraw-Hill Inc., Box 543, Blacklick, OH 43004-0543

Materials of Construction, 1994, 4th edition, R. C. Smith and C. K. Andres, McGraw-Hill Inc., Box 543, Blacklick, OH 43004-0543

Construction Materials and Processes, 1990, Watson, McGraw-Hill Inc., Box 543, Blacklick, OH 43004-0543

Design and Control of Concrete Mixtures, 1988/1990, 13th edition, Portland Cement Association, 5420 Old Orchard Road, Skokie, IL 60077-1083

NRCA Roofing and Waterproofing Manual, 1996, 4th edition, National Roofing Contractor's Association, 10255 W. Higgins, Rd., Suite #600, Rosemont, IL 60018-5607

Recommended Reading for the Contractor's Exams (continued)

Commercial/Industrial Contractor's Exam (continued)

📖 *Principles and Practices of Heavy Construction,* 1993, R. C. Smith & C.K. Andres, Prentice-Hall Co., P. O. Box 11071, Des Moines, IA 50336-1071

📖 *Excavation & Grading Handbook,* 1991, Nicholas E. Capachi, Craftsman Book Company, 6058 Corte del Cedro, P.O. Box 6500, Carlsbad, CA 92018

📖 *Concrete Construction & Estimating,* 1991, Craig Avery, Craftsman Book Company, 6058 Corte del Cedro, P.O. Box 6500, Carlsbad, CA 92018

📖 *Handbook of Rigging,* 1988 4th edition, W. E. Rossnagel, McGraw-Hill Publishing, Inc., Blue Ridge Summit, PA 17294

📖 *Manual of Steel Construction,* 1989, 9th edition, American Institute of Steel Construction, P. O. Box 806276, Chicago, IL 60680-4124

📖 *Construction of Structural Steel Building Frames,* W. G. Rapp, Roger E. Krieger Publishing Co., P. O. Box 9542, Melbourne, FL 32952

📖 *Modern Masonry,* 1991, C. E. Kicklighter, The Goodheart-Willcox Company, Inc., 123 West Taft Drive, South Holland, IL 60473

📖 *Modern Welding,* 1992, Althouse, Turnequist, Bowditch and Bowditch, The Goodheart-Willcox Company, Inc., 123 West Taft Drive, South Holland, IL 60473

Commercial/Industrial/Residential Contractor's Exam

📖 *SBCCI Standard Building Code 1994,* Southern Building Code Congress International, Inc., 900 Montclair Road, Birmingham, AL 35213-1206

📖 *NFPA 70 - National Electrical Code,* 1996 edition, National Fire Protection Association, 1 Batterymarch Park, P. O. Box 9101, Quincy, MA 02269

📖 *Code of Federal Regulations - Title 29, Part 1926 (OSHA),* July 1995, Superintendent of Documents, U.S. Government Printing Office, Washington, DC 20402

📖 *Carpentry and Building Construction,* 1993, Feirer, Hutchings & Feirer, McGraw-Hill Inc., Box 543, Blacklick, OH 43004-0543

📖 *Materials of Construction,* 1994, 4th edition, R. C. Smith and C. K. Andres, McGraw-Hill Inc., Box 543, Blacklick, OH 43004-0543

📖 *Design and Control of Concrete Mixtures,* 1988/1990, 13th edition, Portland Cement Association, 5420 Old Orchard Road, Skokie, IL 60077-1083

📖 *NRCA Roofing and Waterproofing Manual,* 1996, 4th edition, National Roofing Contractor's Association, 10255 W. Higgins, Rd., Suite #600, Rosemont, IL 60018-5607

Recommended Reading for the Contractor's Exams (continued)

Commercial/Industrial/Residential Contractor's Exam (continued)

📖 *Excavation & Grading Handbook,* 1991, Nicholas E. Capachi, Craftsman Book Company, 6058 Corte del Cedro, P.O. Box 6500, Carlsbad, CA 92018

📖 *Concrete Construction & Estimating,* 1991, Craig Avery, Craftsman Book Company, 6058 Corte del Cedro, P.O. Box 6500, Carlsbad, CA 92018

📖 *Principles and Practices of Heavy Construction,* 1993, R. C. Smith & C.K. Andres, Prentice-Hall Co., P. O. Box 11071, Des Moines, IA 50336-1071

📖 *Handbook of Rigging,* 1988 4th edition, W. E. Rossnagel, McGraw-Hill Publishing, Inc., Blue Ridge Summit, PA 17294

📖 *Construction of Structural Steel Building Frames ,* W. G. Rapp, Roger E. Krieger Publishing Co., P. O. Box 9542, Melbourne, FL 32952

📖 *Modern Masonry,* 1991, C. E. Kicklighter, The Goodheart-Willcox Company, Inc., 123 West Taft Drive, South Holland, IL 60473

📖 *Modern Carpentry,* 1992, Willis H. Wagner, The Goodheart-Willcox Company, Inc., 123 West Taft Drive, South Holland, IL 60473

Electrical Contractor's Exam

📖 *NFPA 70 - National Electrical Code,* 1996 edition, National Fire Protection Association, 1 Batterymarch Park, P. O. Box 9101, Quincy, MA 02269

📖 *American Electricians Handbook,* 1996, 13th edition, Croft/Summers, McGraw-Hill Inc., Box 543, Blacklick, OH 43004-0543

📖 *Electrical Review for Electricians,* 9th edition, 1996, J. Morris Trimmer and Charles Pardue, Construction Bookstore, 100 Enterprise Place, Dover, DE 19903-7029

📖 *National Electrical Code Blueprint Reading,* K. L. Gebert, American Technical Publishers, 1155 West 175th Street, Homewood, IL 60403

High Voltage Electrical Contractor's Exam

📖 *NFPA 70 - National Electrical Code,* 1996 edition, National Fire Protection Association, 1 Batterymarch Park, P. O. Box 9101, Quincy, MA 02269

📖 *Cableman's and Lineman's Handbook,* 8th edition, Kurtz and Shoemaker, McGraw-Hill Inc., Box 543, Blacklick, OH 43004-0543

📖 *American Electricians Handbook,* 1996, 13th edition, Croft/Summers, McGraw-Hill Inc., Box 543, Blacklick, OH 43004-0543

Recommended Reading for the Contractor's Exams (continued)

HVAC Refrigeration Contractor's Exam

📖 *Standard Mechanical Code 1994*, Southern Building Code Congress International, Inc., 900 Montclair Road, Birmingham, AL 35213-1206

📖 *Standard Gas Code 1994,* Southern Building Code Congress International, Inc., 900 Montclair Road, Birmingham, AL 35213-1206

📖 *Standard Building Code 1994,* Southern Building Code Congress International, Inc., 900 Montclair Road, Birmingham, AL 35213-1206

📖 *SBCCI Standard Building Code 1994,* Southern Building Code Congress International, Inc., 900 Montclair Road, Birmingham, AL 35213-1206

📖 *Modern Refrigeration and Air Conditioning,* 1992, Althouse, Turnquist and Bracciano, The Goodheart-Willcox Company, Inc., 123 West Taft Drive, South Holland, IL 60473

Industrial Building Contractor's Exam

📖 *SBCCI Standard Building Code 1994,* Southern Building Code Congress International, Inc., 900 Montclair Road, Birmingham, AL 35213-1206

📖 *NFPA 70 - National Electrical Code,* 1996 edition, National Fire Protection Association, 1 Batterymarch Park, P. O. Box 9101, Quincy, MA 02269

📖 *Code of Federal Regulations - Title 29, Part 1926 (OSHA),* July 1995, Superintendent of Documents, U.S. Government Printing Office, Washington, DC 20402

📖 *Construction of Structural Steel Building Frames*, W. G. Rapp, Roger E. Krieger Publishing Co., P. O. Box 9542, Melbourne, FL 32952

📖 *Design and Control of Concrete Mixtures,* 1988/1990, 13th edition, Portland Cement Association, 5420 Old Orchard Road, Skokie, IL 60077-1083

📖 *NRCA Roofing and Waterproofing Manual,* 1996, 4th edition, National Roofing Contractor's Association, 10255 W. Higgins, Rd., Suite #600, Rosemont, IL 60018-5607

📖 *Principles and Practices of Heavy Construction,* 1993, R. C. Smith & C.K. Andres, Prentice-Hall Co., P. O. Box 11071, Des Moines, IA 50336-1071

📖 *Materials of Construction,* 1994, 4th edition, R. C. Smith and C. K. Andres, McGraw-Hill Inc., Box 543, Blacklick, OH 43004-0543

📖 *Handbook of Rigging,* 1988 4th edition, W. E. Rossnagel, McGraw-Hill Publishing, Inc., Blue Ridge Summit, PA 17294

Recommended Reading for the Contractor's Exams (continued)

Industrial Building Contractor's Exam (continued)

Moving the Earth, 3rd edition, 1987, H. L. Nichols, Jr., McGraw-Hill Inc., Box 543, Blacklick, OH 43004-0543

Manual of Steel Construction, 1989, 9th edition, American Institute of Steel Construction, P. O. Box 806276, Chicago, IL 60680-4124

Concrete Construction & Estimating, 1991, Craig Avery, Craftsman Book Company, 6058 Corte del Cedro, P.O. Box 6500, Carlsbad, CA 92018

Modern Welding, 1992, Althouse, Turnequist, Bowditch and Bowditch, The Goodheart-Willcox Company, Inc., 123 West Taft Drive, South Holland, IL 60473

Residential/Small Commercial Contractor's Exam

SBCCI Standard Building Code 1994, Southern Building Code Congress International, Inc., 900 Montclair Road, Birmingham, AL 35213-1206

NFPA 70 - National Electrical Code, 1996 edition, National Fire Protection Association, 1 Batterymarch Park, P. O. Box 9101, Quincy, MA 02269

Code of Federal Regulations - Title 29, Part 1926 (OSHA), July 1995, Superintendent of Documents, U.S. Government Printing Office, Washington, DC 20402

Carpentry and Building Construction, 1993, Feirer, Hutchings & Feirer, McGraw-Hill Inc., Box 543, Blacklick, OH 43004-0543

Design and Control of Concrete Mixtures, 1988/1990, 13th edition, Portland Cement Association, 5420 Old Orchard Road, Skokie, IL 60077-1083

Modern Carpentry, 1992, Willis H. Wagner, The Goodheart-Willcox Company, Inc., 123 West Taft Drive, South Holland, IL 60473

NRCA Roofing and Waterproofing Manual, 1996, 4th edition, National Roofing Contractor's Association, 10255 W. Higgins, Rd., Suite #600, Rosemont, IL 60018-5607

Construction of Structural Steel Building Frames , W. G. Rapp, Roger E. Krieger Publishing Co., P. O. Box 9542, Melbourne, FL 32952

Principles and Practices of Heavy Construction, 1993, R. C. Smith & C.K. Andres, Prentice-Hall Co., P. O. Box 11071, Des Moines, IA 50336-1071

Materials of Construction, 1994, 4th edition, R. C. Smith and C. K. Andres, McGraw-Hill Inc., Box 543, Blacklick, OH 43004-0543

Excavation & Grading Handbook, 1991, Nicholas E. Capachi, Craftsman Book Company, 6058 Corte del Cedro, P.O. Box 6500, Carlsbad, CA 92018

Recommended Reading for the Contractor's Exams (continued)

Residential/Small Commercial Contractor's Exam (continued)

📖 *Manual of Steel Construction,* 1989, 9th edition, American Institute of Steel Construction, P. O. Box 806276, Chicago, IL 60680-4124

📖 *Using Gypsum Board for Walls and Ceilings - GA-201-90,* 1993, Gypsum Association, 810 First Street NE, Washington, DC 20002

📖 *Construction of Structural Steel Building Frames ,* W. G. Rapp, Roger E. Krieger Publishing Co., P. O. Box 9542, Melbourne, FL 32952

Mechanical Contractor's Exam

📖 *Standard Mechanical Code 1994 ,* Southern Building Code Congress International, Inc., 900 Montclair Road, Birmingham, AL 35213-1206

📖 *Standard Gas Code 1994,* Southern Building Code Congress International, Inc., 900 Montclair Road, Birmingham, AL 35213-1206

📖 *Code of Federal Regulations - Title 29, Part 1926 (OSHA),* July 1995, Superintendent of Documents, U.S. Government Printing Office, Washington, DC 20402

📖 *NFPA 13 - Installation of Sprinkler Systems,* 1985, National Fire Protection Association, 1 Batterymarch Park, P. O. Box 9101, Quincy, MA 02269

📖 *NFPA 14 - Installation of Standpipe and Hose Systems,* 1993, National Fire Protection Association, 1 Batterymarch Park, P. O. Box 9101, Quincy, MA 02269

📖 *NFPA 54 - National Fuel Gas Code,* 1996, National Fire Protection Association, 11 Tracey Ave., Avon, MA 02322

📖 *NFPA 58 - Standard for the Storage and Handling of Liquefied Petroleum Gasses,* National Fire Protection Association, 11 Tracey Ave., Avon, MA 02322

📖 *NFPA 90A - Installation of Air Conditioning and Ventilating Systems,* 1993, National Fire Protection Association, 1 Batterymarch Park, Box 9101, Quincy, MA 02269-9101

📖 *ANSI/ASHRAE 15-94 - Safety Code for Mechanical Refrigeration,* 1994 ANSI, 11 West 42nd St., New York, NY 10036

📖 *ASME B31.1 - Power Piping,* 1995, American Society of Mechanical Engineers, 22 Law Drive, P. O. Box 2300, Fairfield, NJ 07007-2300

📖 *ASME B31.8- Gas Transmission and Distribution Piping Systems,* 1995, American Society of Mechanical Engineers, 22 Law Drive, P. O. Box 2300, Fairfield, NJ 07007-2300

Recommended Reading for the Contractor's Exams (continued)

Mechanical Contractor's Exam (continued)

📖 *Refrigeration and Air Conditioning,* 1995, 3rd edition, Air Conditioning and Refrigeration Institute, Prentice Hall, P. O. Box 11071, Des Moines, IA 50336-1071

📖 *Manual N - Load Calculation for Commercial Summer and Winter Air Conditioning,* 1988, 4th edition, Air Conditioning Contractors of America, 1712 New Hampshire Ave., NW, Washington, DC 20009

📖 *Trane Air Conditioning Manual,* 1965, Trane Company, 8929 Western Way, Suite #1, Jacksonville, FL 32256

📖 *Trane Ductulator,* 1976, Trane Company, 8929 Western Way, Suite #1, Jacksonville, FL 32256

📖 *System Design Manual - Part III,* Piping Design, Carrier, 1960, Carrier Air Conditioning Company, Carrier Parkway, P. O. Box 4808, Syracuse, NY 13221

📖 *Pipefitters Handbook,* 1967, 3rd edition, Forest Lindsey, Industrial Press, Inc., 200 Madison Avenue, New York, NY 10016

📖 *Pipe Welding Procedures,* 1973, H. Rampaul, Industrial Press, Inc., 200 Madison Avenue, New York, NY 10016

📖 *Plumbing,* L. V. Ripka, American Technical Publishers, 1155 West 175th Street, Homewood, IL 60403

📖 *Low Pressure Boilers,* 1994, F. M. Steingress, American Technical Publishers, 1155 West 175th Street, Homewood, IL 60403

📖 *High Pressure Boilers,* 1993, F. M. Steingress, American Technical Publishers, 1155 West 175th Street, Homewood, IL 60403

Residential Contractor's Exam

📖 *SBCCI Standard Building Code 1994,* Southern Building Code Congress International, Inc., 900 Montclair Road, Birmingham, AL 35213-1206

📖 *NFPA 70 - National Electrical Code,* 1996 edition, National Fire Protection Association, 1 Batterymarch Park, P. O. Box 9101, Quincy, MA 02269

📖 *Code of Federal Regulations - Title 29, Part 1926 (OSHA),* July 1995, Superintendent of Documents, U.S. Government Printing Office, Washington, DC 20402

📖 *Carpentry and Building Construction,* 1993, Feirer, Hutchings & Feirer, McGraw-Hill Inc., Box 543, Blacklick, OH 43004-0543

📖 *NRCA Roofing and Waterproofing Manual,* 1996, 4th edition, National Roofing Contractor's Association, 10255 W. Higgins, Rd., Suite #600, Rosemont, IL 60018-5607

Recommended Reading for the Contractor's Exams (continued)

Residential Contractor's Exam (continued)

📖 *Design and Control of Concrete Mixtures,* 1988/1990, 13th edition, Portland Cement Association, 5420 Old Orchard Road, Skokie, IL 60077-1083

📖 *Excavation & Grading Handbook,* 1991, Nicholas E. Capachi, Craftsman Book Company, 6058 Corte del Cedro, P.O. Box 6500, Carlsbad, CA 92018

📖 *Modern Carpentry,* 1992, Willis H. Wagner, The Goodheart-Willcox Company, Inc., 123 West Taft Drive, South Holland, IL 60473

📖 *Masonry Skills,* 1990, R. T. Kreh, Sr., Delmar Publishers, P. O. Box 6904, Florence, KY 41022

Plumbing Contractor's Exam

📖 *Standard Plumbing Code 1994,* Southern Building Code Congress International, Inc., 900 Montclair Road, Birmingham, AL 35213-1206

📖 *Standard Gas Code 1994,* Southern Building Code Congress International, Inc., 900 Montclair Road, Birmingham, AL 35213-1206

📖 *Plumbing Technology: Design and Installation,* 1994, 2nd edition, Lee Smith, Delmar Publishers, P. O. Box 6904, Florence, KY 41022

📖 *Plumbing,* L. V. Ripka, American Technical Publishers, 1155 West 175th Street, Homewood, IL 60403

Small Commercial Contractor's Exam

📖 *SBCCI Standard Building Code 1994,* Southern Building Code Congress International, Inc., 900 Montclair Road, Birmingham, AL 35213-1206

📖 *NFPA 70 - National Electrical Code,* 1996 edition, National Fire Protection Association, 1 Batterymarch Park, P. O. Box 9101, Quincy, MA 02269

📖 *Code of Federal Regulations - Title 29, Part 1926 (OSHA),* July 1995, Superintendent of Documents, U.S. Government Printing Office, Washington, DC 20402

📖 *Carpentry and Building Construction,* 1993, Feirer, Hutchings & Feirer, McGraw-Hill Inc., Box 543, Blacklick, OH 43004-0543

📖 *Design and Control of Concrete Mixtures,* 1988/1990, 13th edition, Portland Cement Association, 5420 Old Orchard Road, Skokie, IL 60077-1083

📖 *Excavation & Grading Handbook,* 1991, Nicholas E. Capachi, Craftsman Book Company, 6058 Corte del Cedro, P.O. Box 6500, Carlsbad, CA 92018

Recommended Reading for the Contractor's Exams (continued)

Small Commercial Contractor's Exam (continued)

📖 *NRCA Roofing and Waterproofing Manual,* 1996, 4th edition, National Roofing Contractor's Association, 10255 W. Higgins, Rd., Suite #600, Rosemont, IL 60018-5607

📖 *Gypsum Construction Handbook,* 1992, United States Gypsum Company, 125 South Franklin St., P. O. Box 806278, Chicago, IL 60680-4124

📖 *Principles and Practices of Heavy Construction,* 4th edition, R. C. Smith and C. K. Andres, Prentice-Hall Publishers, 200 Old Tappan, Old Tappan, NJ -07675

📖 *Modern Carpentry,* 1992, Willis H. Wagner, The Goodheart-Willcox Company, Inc., 123 West Taft Drive, South Holland, IL 60473

Fire Protection Sprinkler System

To be a Fire Protection Sprinkler System contractor, you must get a certificate of registration from the Division of Fire Prevention, Permits and Licenses Section. To get an application, you can contact them at:

Contact

Department of Commerce and Insurance
Division of Fire Prevention
Permits and Licenses Section
500 James Robertson Parkway, 3rd floor
Nashville, TN 37243-1159
(615) 741-1322
Fax: (615) 751-1583

To register as a contractor you must certify that you are familiar with the Tennessee Fire Protection Sprinkler System Code Rules and Regulations. If you aren't licensed by the Tennessee Board for Licensing Contractors, you must post a $10,000 surety bond. You also have to provide a list of Responsible Managing Employees that are licensed in Tennessee (see below for information on Responsible Managing Employee).

Fee

Fire protection sprinkler system contractor certificate fees: It will cost you $100 nonrefundable to file an application. The Certificate will cost you $500 and it's good for one year, expiring on June 30.

To qualify as a Responsible Managing Employee of a Fire Protection Sprinkler System contractor, you must be registered in Tennessee as a professional engineer or architect, or complete the National Institute for Certificate in Engineering Technology (NICET) requirements for certification at Level III for fire protection

---

automatic sprinkler systems layout. You must be familiar with the Tennessee Fire Protection Sprinkler System Code Rules and Regulations. And you have to apply as an employee of a Fire Protection Sprinkler System contractor.

Here are the details on the sprinklers and fire protection exam (40 multiple choice questions, two hours):

Subject	Approximate percent of exam
Sprinkler system	45
Wet, dry chemical systems	25
Water supply, storage, distribution	10
General knowledge	10
Flammable knowledge	5
Commercial kitchen equipment	5

Responsible managing employee of a fire protection sprinkler system contractor certificate fees: It will cost you $25 nonrefundable to file an application. The Certificate will cost you $200 and it's good for one year, expiring on June 30.

Recommended Reading for Fire Protection Sprinkler System Exam

NFPA 70 - National Electrical Code, 1996 edition, National Fire Protection Association, 1 Batterymarch Park, P. O. Box 9101, Quincy, MA 02269

NFPA 13 - Installation of Sprinkler Systems, 1985, National Fire Protection Association, 1 Batterymarch Park, P. O. Box 9101, Quincy, MA 02269

NFPA 14 - Installation of Standpipe and Hose Systems, 1993, National Fire Protection Association, 1 Batterymarch Park, P. O. Box 9101, Quincy, MA 02269

NFPA 17 - Dry Chemical Extinguishing Systems, 1990, National Fire Protection Association, 1 Batterymarch Park, P. O. Box 9101, Quincy, MA 02269

NFPA 17A - Wet Chemical Extinguishing Systems, 1990, National Fire Protection Association, 1 Batterymarch Park, P. O. Box 9101, Quincy, MA 02269

Code of Federal Regulations - Title 29, Part 1926 (OSHA), July 1995, Superintendent of Documents, U.S. Government Printing Office, Washington, DC 20402

Department of Transportation (DOT)

You have to be prequalified to bid on Tennessee Department of Transportation projects. To get an application, contact:

Tennessee Department of Transportation
James K. Polk Building, Suite 700
Nashville, TN 37243-0326
(615) 741-3408
Fax: (615) 741-0782

Some of the details the Department will ask you about are your organization, personnel, financial condition, equipment, and experience. You'll be asked what type of work you want to be qualified for.

Prequalification is good for one year, with a three-month grace period at the end.

Out-of-State Corporations

Out-of-state corporations must get a Certificate of Authority to do business in Tennessee from the Tennessee Secretary of State. To apply for this certificate, contact:

Department of the Secretary of State
Division of Certification
James K. Polk Building, Suite 1800
Nashville, TN 37243
(615) 741-2286
Fax: (615) 741-7310

Texas

Only specialty contractors, including HVAC, fire sprinkler systems, plumbing, and well drilling/pump installation specialists, need to be licensed in Texas.

HVAC Contractor's Licenses

To do HVAC work in Texas you need to be licensed. To get an application, contact:

Contact

Texas Department of Licensing and Regulation
E. O. Thompson State Office Building
P. O. Box 12157
Austin, TX 78711
(512) 463-6599
(800) 803-9202 (in Texas only)
Fax: (512) 475-2871

The Department issues two types of licenses — Class A and Class B. The Class A license lets you work on any size HVAC equipment. The Class B license limits you to 25 tons of cooling and 1.5 million Btu of heating. For either the Class A or Class B license, you also need an endorsement for environmental air conditioning, commercial refrigeration and process cooling and heating, or both.

You have to pass an exam for each type of endorsement. To take an exam you need at least three years of practical work experience in the preceding five years. If you have a degree in air conditioning engineering, refrigeration engineering, or mechanical engineering from a Department-approved school, you can use it for two years of the work experience requirement.

The exams are open book with true/false and multiple choice questions. The Class A exams have 100 questions and the Class B exams have 50 questions. Here's a list of the subjects on each exam and the number of questions on each subject for Class A and Class B:

Environmental air conditioning exam:

Subject	Class A exam	Class B exam
Uniform Mechanical or *Standard Mechanical Code*	30	16
Boilers	10	3
Design and construction	20	4
Maintenance, service, repair	20	15
Rules, regulations	10	5
Refrigeration cycle	10	7

Commercial refrigeration and process cooling and heating exam:

Subject	Class A exam	Class B exam
Uniform Mechanical or *Standard Mechanical Code*	20	10
Boiler laws, rules	5	2
Air conditioning and refrigeration laws, rules	10	5
Boilers and process heating	10	5
General refrigeration, process cooling	20	10
Refrigeration, process cooling maintenance, repair	13	7
Refrigeration, process cooling controls	12	6
Refrigeration safety	5	2
Piping	5	3

After you pass the exams you have these insurance requirements for a license:

- Class A license: $300,000 per occurrence for property damage and bodily injury, $300,000 aggregate for property damage and bodily injury, and $300,00 aggregate for products and completed operations
- Class B license: $100,000 per occurrence for property damage and bodily injury, $100,000 aggregate for property damage and bodily injury, and $100,00 aggregate for products and completed operations

If you have a valid HVAC license in Georgia or South Carolina you can use it to get a Texas license.

HVAC contractor's license fees: It will cost you $50 to file an application. Each exam costs $50. A license costs $125 and it's good for three years.

Recommended Reading for the HVAC Exams

Environmental Air Conditioning Exam

- *Uniform Mechanical Code 1994,* International Conference of Building Officials, 9300 Jollyville Road, Suite 101, Austin, TX 78759-7455

- *Standard Gas Code 1994,* Southern Building Code Congress International, Inc., 9420 Research Blvd., Echelon III, Suite 150, Austin, TX 78759

- *Standard Mechanical Code 1994,* Southern Building Code Congress International, Inc., 9420 Research Blvd., Echelon III, Suite 150, Austin, TX 78759

- *Texas Boiler Law, Health and Safety Code, Chapter 755, and Administrative Rules,* Texas Dept. of Licensing and Regulation, P. O. Box 12157, Austin, TX 78711

- *Modern Refrigeration and Air Conditioning,* 1992, Althouse, Turnquist and Bracciano, The Goodheart-Willcox Company, Inc., 123 West Taft Drive, South Holland, IL 60473

Commercial Refrigeration Exam

- *Refrigeration and Air Conditioning,* 1995, 3rd edition, Air Conditioning and Refrigeration Institute, Prentice Hall, P. O. Box 11071, Des Moines, IA 50336-1071

- *High Pressure Boilers,* 1993, F. M. Steingress, American Technical Publishers, 1155 West 175th Street, Homewood, IL 60403

- *Low Pressure Boilers,* 1994, F. M. Steingress, American Technical Publishers, 1155 West 175th Street, Homewood, IL 60403

- *ANSI/ASHRAE Standard 15-94 Safety Code for Mechanical Refrigeration,* American National Standards Institute, 11 West 42nd Street, New York, NY 10036

- *Pipefitters Handbook,* 1967, 3rd edition, Forest Lindsey, Industrial Press, Inc., 200 Madison Avenue, New York, NY 10016

Fire Sprinkler Systems Contractor's License

To work as a fire sprinkler systems contractor in Texas you must get a certificate of registration for your business. To get an application, contact:

Contact

State Fire Marshall's Office
P. O. Box 149221
Austin, TX 78714-9221
(512) 918-7222
Fax: (512) 918-7107

The Fire Marshall issues three types of certificates — general, dwelling, and underground water supply piping. Here are the general requirements you must have to get a certificate:

- $100,000 insurance per occurrence with $300,000 total coverage
- authority to do business in Texas from the Texas Secretary of State or the County Clerk
- letter of good standing with the Texas Comptroller (franchise tax)

You must also employ at least one full-time person you designate as a Responsible Managing Employee (RME) who is licensed by the Fire Marshall. To get a RME license, you must pass an exam on the statutes and rules in the Texas Insurance Code Article 5.43-3 — Fire Protection Sprinkler Systems. The Fire Marshall's Office gives this exam.

Unless you're registered as a professional engineer in Texas you must also pass a course given by the National Institute for Certification in Engineering Technologies (NICET) which includes an exam. For a RME general license you must pass the NICET III course. For the RME dwelling license you need to pass a NICET II course and a Fire Marshall's Office exam on the NFPA 13D — Installation of Sprinkler Systems in One and Two Family Dwellings and Mobile Homes. For a RME underground piping license you don't need any NICET course but you'll have to pass a Fire Marshall's Office exam on NFPA 24 — Private Fire Service Mains and Their Appurtenances.

Exams given by the Fire Marshall's Office are closed book with 25 to 50 true/false or multiple choice questions. All exams last three hours.

To get information on NICET courses and exams, you can contact them at:

Contact

National Institute for Certification in Engineering Technologies
1420 King Street
Alexandria, VA 22314
(800) 787-0034

Fee

Fire sprinkler systems contractor's license fees: You'll have to pay $50 nonrefundable to take any exam given by the Fire Marshall's Office. It costs $50 to apply for a certificate for your company or a RME license. A general certificate costs $900 and a dwelling or underground piping certificate costs $300. A RME general license costs $175 and a dwelling or underground license is $100. A certificate or license is good for two years.

If you have a valid license from another state with requirements generally the same as those in Texas, the Fire Marshall's Office may waive any of its license requirements.

Plumber's Licenses

You need a license to do plumbing work in Texas. To get an application, contact:

Texas State Board of Plumbing Examiners
929 East 41st Street
P. O. Box 4200
Austin, TX 78765-4200
(512) 458-2145
(800) 845-6584 (in Texas only)
Fax: (512) 450-0637

You'll have to pass an exam to get a plumbing license. To qualify for the master exam you must have held a journeyman license in Texas for two years. If you have a journeyman or master license in another state you can get a Texas journeyman license directly. Then you don't have to pass the two-year Texas journeyman license period. Texas doesn't have any reciprocity agreements with other states.

The master exam lasts two days. It has a written part and a hands-on part. The hands-on part of the exam will ask you to install sanitary plumbing in a three-story miniature commercial structure. This will test your ability to lay out and install a sanitary waste and vent system. You'll have to make a material take-off for the rough-in of the waste and vent systems for the fixture shown on the miniature commercial structure.

The written part has multiple choice questions on the following topics:

Waste and vent systems

Natural gas systems

Water systems

Simple physics

Trade-related math problems

Terms and definitions

OSHA, ADA, LPG

On-site sewerage facilities

Cross-connections created by
   back siphonage and backflow

All questions come from the three codes used in Texas — the *Standard Plumbing Code*, the *Uniform Plumbing Code*, and the *National Standard Plumbing Code*. There will also be 25 multiple choice questions on licensing law and Board rules and regulations.

Plumber's license fees: It will cost you $150 to take the master exam. The fee you pay for a license and its duration will depend on when Board issues it to you.

To qualify to take the journeyman exam, you must have at least 8,000 hours of work experience in the trade or a combination of work experience and technical training that equals 8,000 hours. If you have a journeyman or master license in another state, you qualify to take the Texas journeyman exam.

The journeyman exam has a written part and a hands-on part. The written part of the exam has multiple choice questions on the following topics:

Waste and vent systems	Water systems
Natural gas systems	Simple physics
Cross-connections created by back siphonage and backflow	Terms and definition

All questions come from the three codes used in Texas — the *Standard Plumbing Code*, the *Uniform Plumbing Code*, the *National Standard Plumbing Code,* and the *Plumbing License Law and Board Rules*. For more information on these sources see the discussion on the master plumbing exam references. There will also be ten multiple choice questions on licensing law and Board rules and regulations.

The hands-on part of the exam has two parts. The first part will be on:

- gas burner adjustment
- identification of plastic, copper and steel fittings
- using hand tools and brazing equipment to sweat and braze copper tube, threaded steel pipe, PVC plastic pipe, copper flare and cast iron pipe with compressing no-hub couplings

You need to know about pipe formulas and proper methods to prepare materials to join them together for this part of the exam.

The second part of the hands-on exam will ask you to install sanitary plumbing in a two-story miniature residential structure. This will test your ability to lay out and install a sanitary waste and vent system. You'll have to make a material take-off for the rough-in of the waste and vent systems for the fixture shown on the miniature residential structure.

Journeyman plumber's exam fee: It will cost you $25 to take the journeyman exam. The fee you pay for a license and its duration will depend on when Board issues it to you.

Recommended Reading for the Plumbing Exam

- *Construction Standards for On-Site Sewerage Facilities,* Texas State Board of Plumbing Examiners

- *ADA Accessibility Guidelines,* Secretary of State, Texas Register Division, P. O. Box 13824, Austin, TX 78711

- *The Plumbing License Law and Board Rules,* Texas State Board of Plumbing Examiners

- *Liquefied Petroleum Gas Safety Rules,* May 1994, Thomas Petru, Texas Railroad Commission Division, P. O. Box 12967, Austin, TX 78711-2967

Recommended Reading for the Plumbing Exam (continued)

📖 *Code of Federal Regulations - Title 29, Part 1926 (OSHA)*, July 1995, Superintendent of Documents, U.S. Government Printing Office, Washington, DC 20402

📖 *National Plumbing Code*, Associated Plumbing, Heating & Cooling Contractors of Texas, 2201 N. Lamar, Suite 102, Austin, TX 78705

📖 *Uniform Plumbing Code*, International Association of Plumbing and Mechanical Officials, 20001 Walnut Drive South, Walnut, CA

📖 *Standard Plumbing Code*, Southern Building Code Congress International, Inc., 9420 Research Blvd., Suite 150, Austin, TX 78759

Well Drilling/Pump Installation License

To do well drilling or pump installation work in Texas you must get a license. To get an application, contact:

Contact

Texas Department of Licensing and Regulation
P. O. Box 12157
Austin, TX 78711
(512) 463-7880
Fax: (512) 463-8616

The Department issues four types of well drilling license: monitor well, dewatering well, injection well, and water well. The water well licenses lets you work on any of the other types of well drilling.

The Department issues these five types of pump installer licenses:

Windmills, hand pump, pump jacks   Line-shaft turbine pumps

Pumps (submersible 5-hp and over)   Master

Pumps (fractional to 5-hp)

The master license lets you work on the other four types of pump installations.

On your license application you'll have to give the Department complete information on the following:

● at least two years of well drilling or pump installation work experience
● letters of reference on your competence from four persons who have been licensed as drillers or pump installers in Texas for at least two years
● letter of reference from your banker
● letters of reference from two satisfied customers

You'll also have to pass an exam the Department has developed to test your abilities in well drilling and/or pump installation to get a license.

You also need to prove you've lived in Texas for 90 days before you apply for a license and that you intend to reside permanently in the state. If you have a valid license from another state with requirements generally the same as those in Texas, the Department may waive any of the requirements it has for a license. The state you're licensed in must also extend the same privilege to drillers or pump installers licensed by the Department in Texas.

Well drilling/pump installation license fees: It will cost you $100 to apply for a well driller or pump installer license and take an exam. A combination driller/installer application and exam costs $200. A license also costs $100. All licenses expire August 31 each year.

Department of Transportation (DOT)

You have to be prequalified to bid on Texas Department of Transportation projects. To get an application contact:

Texas Department of Transportation
125 East 11th Street
Austin, TX 78701-2483
(512) 416-2540
Fax: (512) 416-2538

Some of the details the Department will ask you about are your organization, personnel, financial condition, equipment, and experience. You'll be asked what type of work you want to be qualified for. Here are the types of work the Department uses:

Asphalt	Cleaning, sweeping highways
Traffic control devices	Landscaping
Concrete paving, incidentals	Building construction
Other	Major structures
Earthwork, base & subbase	Rest, picnic area maintenance
Guardrail repair	Material supplier
Engineering	Hazardous material
Debris clearing, removal	Minor structure, miscellaneous concrete
Fencing	Underwater inspection
Mowing	Painting, striping
Hauling	Stream channel restoration
Litter pickup, disposal	Rest areas (construction)
Lighting, signal maintenance	Pavement markers

You also need to list the districts in Texas that you're willing to work in. Prequalification is good for one year.

Out-of-State Corporations

Out-of-state corporations must get a Certificate of Authority to do business in Texas from the Texas Secretary of State. To apply for this certificate, contact:

Contact

Corporations Section
Office of the Secretary of State
P. O. Box 13697
Austin, TX 78711-3697
(512) 463-5555
Fax: (512) 463-5709

Utah

To do construction work in Utah you need a license from the Division of Occupational and Professional Licensing. To get an application, contact:

Contact

Division of Occupational & Professional Licensing
160 East 300 South, 4th floor
P. O. Box 146741
Salt Lake City, UT 84145-6741
(801) 530-6436
Fax: (801) 530-6511

Construction Contractor's Licenses

The Division issues these primary contractor classifications:

General engineering
General building
Residential, small commercial
Factory built house set-up
General engineering trades instructor
General building trades instructor
Electrical trades instructor
Plumbing trades instructor
Mechanical trades instructor
General electrical*
General plumbing*
Carpentry
Glass, glazing
Metal, vinyl siding
General painting
Insulation
General concrete

Excavation, grading
Steel erection
Landscaping
Sheet metal
HVAC
Refrigeration
Fire suppression systems
Swimming pool, spa
Sewer and water pipeline
Asphalt paving
Pipeline, conduit
General fencing, guardrail
General masonry
Sign installation
Mechanical insulation
Wrecking, demolition
Petroleum systems

General drywall, stucco, plastering

General roofing

Metal firebox, fuel burning stove installer

Pier, foundations

Wood flooring

*Must obtain separate license. See below.

The Division issues these secondary contractor classifications:

Residential, small commercial nonstructural remodeling & repair

Residential electrical

Boiler installation

Irrigation sprinkling

Industrial piping

Water conditioning equipment

Solar energy systems

Residential sewer connection, septic tank

Residential plumbing

Cabinet, millwork installation

Rain gutter installation

Concrete form setting, shoring

Gunite, pressure grouting

Cementations coating systems, resurfacing, sealing

Plastering, stucco

Ceiling grid systems, ceiling tile, panel systems

Lightweight metal, non-bearing wall partitions

Drywall

Single ply, specialty coating

Build-up roofing

Shingle, shake roofing

Tile roofing

Metal roofing

Stone masonry

Terrazzo

Marble, tile, ceramic

Cultured marble

Steel reinforcing

Metal building erection

Structural stud erection

Refrigerated air conditioning

Evaporative cooling

Warm air heating

Residential fencing

Nonelectric outdoor advertising sign

To qualify for any license, you must:

- provide proof of at least $100,000 for each incident and $300,000 total liability insurance
- provide proof of workers' compensation insurance
- provide registration with Utah State Tax Commission
- provide proof of registration with Utah Department of Employment Security
- provide proof of registration with the Internal Revenue Service
- provide proof of registration with Utah Division of Corporations
- provide proof of DBA registration with Utah Division of Corporations
- submit CPA complied, reviewed, or audited financial statement
- submit three credit reports for all key company personnel

Some licenses also require you to pay $195 to the Residence Lien Recovery Fund.

Construction Contractor's Exams

You must also pass a business and law exam and a trade exam. To qualify for the exams for most trades, you must have at least two years of full-time related work experience in any trade you want to get a license for. But for these six trades, you need four years of full-time related work experience:

General roofing	General masonry
Steel erection	Refrigeration
Heating, ventilating, air conditioning	Fire suppression systems

If you want to take the exam for any of these trades, you need four years of full-time related work experience with two years as a supervisor or manager:

General engineering	General building
Residential, small commercial building	

The exams are given by the National Assessment Institute (NAI). For information on an exam, contact:

Contact

National Assessment Institute, Inc.
560 East 200 South, Suite 300
Salt Lake City, UT 84102
(801) 355-5009

The business and law exam is open book with multiple choice questions on state and federal laws and business financial and accounting procedures. Here's information on the subjects on the exam and the approximate percentage of questions on each subject on the exam:

Subject	Approximate percent of exam
Licensing	20
Estimating, bidding	12
Tax laws	10
Labor laws	10
Financial management	8
Project management	8
Contract management	8
Risk management	8
Safety	6
Business planning, organization	5
Lien law	5

The reference book for this exam is the *Utah Contractor's Reference Manual.* You can buy this book from NAI for $45 plus tax, shipping, and handling. NAI also sells a practice business and law exam.

All trade exam are open book using code books and OSHA construction standards. Questions are on:

- how to bid and manage construction
- how to read and interpret codes and regulations
- trade materials

You can get a content outline for most of the trades from NAI. They also sell practice general building and residential exams.

Construction contractor's license fees: It will cost you $200 nonrefundable to file an application for a license. The business and law exam will cost you $70. Each trade exam costs $75. The license is free and it's good for two years, expiring July 31 of the odd-numbered years.

If you have a valid license in any of the following states, you can apply for a Utah license by endorsement:

Alabama	Hawaii	North Carolina
Arizona	Louisiana	South Carolina
Arkansas	Michigan	Tennessee
California	Mississippi	Virginia
Florida	Nevada	West Virginia
Georgia	New Mexico	

Electrician's Licenses

To do electrical work in Utah you must be licensed. To get an application contact:

Division of Occupational & Professional Licensing
160 East 300 South
P. O. Box 45805
Salt Lake City, UT 84145-0805
(801) 530-6628
Fax: (801) 530-6511

The Division issues master, master residential, journeyman, and journeyman residential licenses. You must pass an exam to get a license.

To qualify for the master exam you must have one of the following:

- a bachelor's degree from a Division-approved electrical engineering program and one year of full-time Division-approved work experience
- an associate's degree from a Division-approved applied science program and two years of full-time Division-approved work experience
- eight years of full-time Division-approved work experience

To qualify for the residential master exam you must have two years of full-time Division-approved work experience as a licensed residential journeyman electrician.

To qualify for the journeyman exam you must have four years of full-time apprentice training or six years of full-time Division-approved work experience. To qualify for the residential journeyman exam you must have two years of full-time apprentice training and two years of full-time Division-approved work experience or four years of full-time Division-approved work experience.

Electrical Exams

The electrical exams are given by National Assessment Institute (NAI). For information on the exams contact:

National Assessment Institute, Inc.
560 East 200 South, Suite 300
Salt Lake City, UT 84102
(801) 355-5009

Each electrical exam has three written parts. Part I is open book with 40 multiple choice questions on electrical theory and calculations. It lasts two hours. Part II is closed book with 40 multiple choice questions on basic electrical theory, calculations, and general knowledge. It lasts two hours. Part III is a practical exam with 30 multiple choice questions on using conduits, switching, motors and controls, transformers, troubleshooting, and general knowledge. It lasts one hour.

Here's information on the subjects on Parts I and II with the approximate percentage of questions on each subject:

	Percent of exam			
Subject	Master	Master residential	Journeyman	Journeyman residential
Ground and bonding	11	12	11	12
Services, feeders, branch circuits, overcurrent protection	11	14	12	14
Raceways and enclosure	10	8	14	8
Conductors	8	14	10	14
Motors and controls	10	3	9	3
Utilization, general use equipment	9	14	10	14
Special occupancies, equipment	10	7	10	7
General knowledge	24	22	17	22
Low voltage circuits	5	4	5	4
State laws and rules	2	2	2	2

Here's the same information for Part III of the electrical exams:

| | Percent of exam | | | |
Subject	Master	Master residential	Journeyman	Journeyman residential
Conduit	9	7	17	7
Switching	11	21	18	21
Motors and controls	25	3	13	3
Transformers	22	3	12	3
General knowledge	15	46	24	46
Troubleshooting	18	20	16	20

If you have a master or journeyman license from any of the states listed below, you can use it to qualify for a Utah licensing exam. Here are the states the Division recognizes:

Alaska	Idaho	New Hampshire
Arkansas	Massachusetts	North Dakota
Colorado	Michigan	Oklahoma
Connecticut	Minnesota	South Dakota
District of Columbia	Montana	Washington

If you have a master or journeyman license from any other state, you must prove you got the license through a procedure at least equal to what the Division uses to license electricians in Utah. Then you can use it to qualify for a Utah licensing exam.

Electrician's license fees: It costs $100 nonrefundable to file an application for a license. The exam will cost you $105. Your first license is free and it's good for two years, expiring July 31 of the even-numbered years.

Recommended Reading for the Electrical Exams

📖 *NFPA 70 - National Electrical Code*, 1996 edition, National Fire Protection Association, 1 Batterymarch Park, P. O. Box 9101, Quincy, MA 02269

📖 *Construction Trades Licensing Act*

📖 *Rules of the Electricians Licensing Board*

📖 *Model Energy Code*, CABO

📖 *Alternating Current Fundamentals*, Duff and Herman, Delmar Publishers, P. O. Box 6904, Florence, KY 41022

Recommended Reading for the Electrical Exams (continued)

📖 *Direct Current Fundamentals,* Loper and Tedsen, Delmar Publishers, P. O. Box 6904, Florence, KY 41022

📖 *Industrial Motor Control Fundamentals,* Herman and Alerich, Delmar Publishers, P. O. Box 6904, Florence, KY 41022

📖 *National Electrical Code Blueprint Reading,* K. L. Gebert, American Technical Publishers, 1155 West 175th Street, Homewood, IL 60403

📖 *American Electricians Handbook,* 1996, 13th edition, Croft/Summers, McGraw-Hill Inc., Box 543, Blacklick, OH 43004-0543

📖 *Guide to the National Electrical Code,* Harman and Allen, Prentice Hall, Inc., 200 Old Tappan, Old Tappan, NJ -07675

You can get these books from:

Builders' Book Depot
1033 East Jefferson, Suite 500
Phoenix, AZ 85034
(800) 284-3434

Plumber's Licenses

To do plumbing work in Utah you must be licensed. To get an application, contact:

Contact

Division of Occupational & Professional Licensing
160 East 300 South
P. O. Box 45805
Salt Lake City, UT 84145-0805
(801) 530-6628
Fax: (801) 530-6511

The Division issues journeyman and residential journeyman licenses. You must pass an exam for either license. To qualify for the journeyman exam you need four years of full-time apprentice training or eight years of full-time Division-approved work experience. Here are the types of work and approximate hours the Division wants you to have:

Work	Approximate hours
Using hand tools, equipment, pipe machinery	150
Installing piping for waste, soil, sewer vent, and leader lines	2,200
Installing hot, cold water for domestic use	1,600
Installing, setting plumbing appliances, fixtures	1,600

Work	Approximate hours
Maintenance, repair plumbing	800
General process & industrial pipe work	800
Installing sheet lead and solder work	550
Gas and service piping	500
Welding	100
Service, maintenance of gas controls, equipment	200

To qualify for the residential journeyman exam you must have three years of full-time apprentice training or six years of full-time Division-approved work experience. Here are the types of work and approximate hours the Division wants you to have:

Work	Approximate hours
Using hand tools, equipment, pipe machinery	100
Installing piping for waste, soil, sewer vent, and leader lines	1,800
Installing hot, cold water for domestic use	1,400
Installing, setting plumbing appliances, fixtures	1,200
Installing sheet lead and solder work	550
Gas and service piping	500

Plumbing Exams

The exams are given by National Assessment Institute (NAI). For information on the exams contact:

Contact

National Assessment Institute, Inc.
560 East 200 South, Suite 300
Salt Lake City, UT 84102
(801) 355-5009

Each plumbing exam has three parts. Part I is open book with 30 multiple choice questions and lasts two hours. You can use these materials for the Part I exam: *Uniform Plumbing Code, Recommended Good Practices for Gas Piping, Appliance Installations and Venting, Mathematics for Plumbers and Pipefitters*, and *The Plumbers Handbook*. See next page for complete reference listing of these materials. Part II is closed book with 70 multiple choice questions and lasts one and one-half hours.

Part III of the exam is a practical exam. For the journeyman license you'll have to complete a copper brazing project, a plastic assembly project, and a copper assembly project in three and one-half hours. For the residential journeyman license you have two and one-half hours to complete a copper solder project and a plastic assembly project.

If you have a journeyman license from another state you can use it to get a Utah license by endorsement. You must prove you got the license through a procedure at least equal to what the Division uses to license plumbers in Utah.

Fee

Plumber's license fees: It will cost you $100 nonrefundable to file an application for a license. The exam will cost you $105. Your first license is free and it's good for two years, expiring July 31 of the even-numbered years.

Recommended Reading for the Plumbing Exams

📖 *Uniform Plumbing Code,* 1991 edition, International Association of Plumbing and Mechanical Officials, 20001 Walnut Drive South, Walnut, CA

📖 *Construction Trades Licensing Act, Rules of the Plumbers Licensing Board,* current edition, Utah Department of Commerce

📖 *Recommended Good Practices for Gas Piping, Appliance Installations, and Venting,* Mountain Fuel Supply Co.

📖 *Mathematics for Plumbers and Pipefitters,* 1996, D'Archangelo, D'Archangelo, and Guest, Delmar ITP, P. O. Box 15015, Albany, NY 12212

📖 *The Plumbers Handbook,* 8th edition, Joseph P. Almond, MacMillan Publishing Co.

📖 *Code of Federal Regulations - Title 29, Part 1926 (OSHA),* July 1995, Superintendent of Documents, U.S. Government Printing Office, Washington, DC 20402

You can get these books from:

Builders' Book Depot
1033 East Jefferson, Suite 500
Phoenix, AZ 85034
(800) 284-3434

Department of Transportation (DOT)

To bid on Utah Department of Transportation projects you have to be prequalified by the Department. To get an application for prequalification contact:

Contact

Utah Department of Transportation
Prequalification Board
4501 South 2700 West
Salt Lake City, UT 84119-5998
(801) 965-4103
Fax: (801) 965-4403

Some of the details the Department will ask you about are your organization, personnel, financial condition, equipment, and experience. You'll be asked what type of work you want to be qualified for. Here are the major types of work the Department uses:

General (all types of work)	Surfacing - plantmix asphalt
Light grading	Surfacing - Portland cement concrete
Heavy grading	Surfacing - miscellaneous
Surfacing - crushed gravel or stone	Bridges, culverts
Surfacing - roadmix asphalt	Miscellaneous

You need a Utah contractor's license to bid on Department of Transportation projects. Prequalification is good for 18 months.

Out-of-State Corporations

Out-of-state corporations must get a Certificate of Authority to do business in Utah from the Utah Secretary of State. To apply for this certificate contact:

Contact

Utah Secretary of State
160 East 300 South, 1st floor
Salt Lake City, UT 84111
(801) 530-4849
Fax: (801) 530-6438

Vermont

In Vermont, construction contractors need to be certified to do asbestos or lead abatement, and licensed to do electrical or plumbing work.

Asbestos Abatement Certifications

Contact

To do asbestos abatement work in Vermont your business must be certified by the Vermont Department of Health. All employees of an asbestos abatement business must also be certified. To get an application, contact:

Asbestos & Lead Regulator Program
Department of Health
108 Cherry Street
P. O. Box 70
Burlington, VT 05402
(802) 863-7231 / (800) 439-8550 (Vermont only)
Fax (802) 863-7425
E-mail: SConger@VDH.VAX.VDH.STATE.VT.US

The Department issues certificates to abatement contractors and consulting contractors. To get an abatement contractor certification you must have:

- two years experience in asbestos abatement projects or three years experience in general contracting
- completed a Department-approved training course in accordance with EPA Asbestos Model Accreditation Plan, 40 CFR Part 763, Subpart E Appendix C
- have set up Department-approved worker protection programs
- certify you don't owe any taxes or child support

For a consulting contractor certification you need to:

- employ only individuals that are certified by the Department or are qualified to do so
- have set up Department-approved worker protection programs.
- certify you don't owe any taxes or child support

Abatement contractor and consulting contractor certificate fees: Certification will cost you $500 and it's good for one year. If your business has been certified in another state, you can petition the Department to be certified in Vermont without repeating the training the Department requires. You still must pay the certification fee.

Any individual working in asbestos abatement must be certified as a worker, supervisor, inspector, inspector management planner, project monitor or project designer. Here's information on each classification and its training and/or experience requirements. You must also certify you don't owe any taxes or child support.

Worker

Required department-approved course: Initial training

Required education and/or asbestos work experience: None

Supervisor

Required department-approved course: Initial training

Required education and/or asbestos work experience: Four months of work experience

Inspector

Required department-approved course: Inspector training

Required education and/or asbestos work experience: One of the following: 1. high school diploma or equivalent and one year of work experience; 2. two years college; 3. three years of experience in engineering or industrial hygiene

Inspector management planner

Required department-approved course: Inspector/management planner training

Required education and/or asbestos work experience: One of the following: 1. one year work experience; 2. two years college and one year work experience; 3. bachelor's degree; 4. three years of work experience in engineering or industrial hygiene

Project monitor

Required department-approved course: Contractor/supervisor training and air sampling for project monitors training

Required education and/or asbestos work experience: One of the following: 1. American Board of Industrial Hygiene Certified Industrial Hygienist status or Registered Professional Engineer or Architect and three months of work experience; 2. bachelor's degree and six months of work experience; 3. two years college and one year work experience; 4. high school diploma or equivalent and four years work experience in engineering or industrial hygiene and one year of work experience in asbestos abatement.

Project designer

Required department-approved course: Project designers or contractor/supervisor training

Required education and/or asbestos work experience: One of the following: 1. American Board of Industrial Hygiene Certified Industrial Hygienist status or Registered Professional Engineer or Architect and six months of work experience; 2. bachelor's degree and one year of work experience; 3. associates degree and two years work experience; 4. high school diploma or equivalent and four years of work experience in engineering or industrial hygiene and one year of work experience in asbestos abatement.

Fee

Asbestos abatement certificate fees: It will cost you $50 for a worker certificate, $100 for a supervisor certificate, and $150 for any other certificate. A certificate is good for one year. If you have been certified in another state, you can petition the Department to be certified in Vermont without repeating the training or work experience the Department requires. You still must pay the certification fee.

Lead Abatement Certifications

To do lead abatement work in Vermont your business must be certified by the Vermont Department of Health. All employees of a lead abatement business must also be certified. To get an application contact:

Contact

Asbestos & Lead Regulator Program
Department of Health
108 Cherry Street
P. O. Box 70
Burlington, VT 05402
(802) 863-7231 / (800) 439-8550 (Vermont only)
Fax (802) 863-7425
E-mail: SConger@VDH.VAX.VDH.STATE.VT.US

The Department issues lead abatement company certificates in these categories:

- lead abatement contractors
    1. target housing and public buildings
    2. superstructures and commercial buildings
- lead consulting contractors

To get a certificate for your company you must:

- have set up Department-approved worker protection programs
- certify you don't owe any taxes or child support
- employ only individuals that are certified by the Department

Lead abatement company certificate fees: It will cost you $500 for a certificate and it's good for one year. If your business has been certified in another state, you can petition the Department to be certified in Vermont without repeating the training the Department requires. You still must pay the certification fee.

Any individual working in lead abatement must be certified as at least one of the following:

Worker

  1. target housing and public buildings
  2. superstructures and commercial buildings

Supervisor

  1. target housing and public buildings
  2. superstructures and commercial buildings

Inspector technician
Inspector/risk assessor
Project designer

Here's information on each classification and its training and/or experience requirements. You must also certify you don't owe any taxes or child support.

Worker

Required Department-approved course: Worker training

Required education and/or lead work experience: None

Supervisor

Required Department-approved course: Supervisor training

Required education and/or lead work experience: One year experience as a lead worker or two years experience in related field or building trades

Inspector technician

Required Department-approved course: Inspector training

Required education and/or lead work experience: None

Inspector/risk assessor

Required Department-approved course: Inspector/risk assessor training

Required education and/or lead work experience: One year work experience or 25 inspections over at least three months as a licensed lead inspector technician and one of the following: 1. bachelor's degree in related field and one year work experience; 2. certification as an industrial hygienist, engineer, architect; 3. high school diploma or equivalent and two years of work experience

Project designer

Required Department-approved course: Project designer training

Required education and/or lead work experience: One of the following: 1. American Board of Industrial Hygiene Certified Industrial Hygienist status or Registered Professional Engineer or Architect and six months lead abatement experience or one year experience as designer for asbestos or radon projects; 2. bachelor's degree and one year lead abatement experience or one year experience as designer for asbestos or radon projects; 3. associates degree and two years of experience in engineering or industrial hygiene and one year lead abatement experience or one year experience as designer for asbestos or radon projects; 4. high school diploma or equivalent and four years experience in engineering or industrial hygiene and one year lead abatement experience or one year experience as designer for asbestos or radon projects

Fee

Lead abatement individual certificate fees: It will cost you $50 for a worker certificate, $100 for a supervisor certificate, and $150 for any other certificate. A certificate is good for one year. If you've been certified in another state, you can petition the Department to be certified in Vermont without repeating the training or work experience the Department requires. You still must pay the certification fee.

Electrician's Licenses

Unless you do electrical work just on residential duplexes or a few other special situations, you'll need a license to work in Vermont. To get information on these exceptions and an application contact:

Contact

State Electricians Licensing Board
Department of Labor and Industry
National Life Bldg., Drawer 20
Montpelier, VT 05620-3401
(802) 828-2107

The Board issues master, journeyman, and Type-S journeyman licenses. Type-S classifications are:

Automatic gas or oil heating	Outdoor advertising
Refrigeration or air conditioning	Appliance and motor repair
Well pumps	Gas pumps and bulk plants
Household fire alarm systems	Commercial fire alarm system
Lightning rod installation	Elevator

The Board will review your application and if you're eligible they'll send you information on the exam they require for a license. For any license, you must certify that you don't owe any taxes or child support.

To qualify for the master license you need two years of experience as a journeyman, Board-approved equivalent experience, or have completed the journeyman's requirements and 4,000 hours work experience. For a journeyman's license you need to complete a Board-approved apprenticeship program of 576 hours and 8,000 hours of experience under a licensed electrician. You may be able to get the Board to accept military or out-of-state work experience for these requirements. For a Type-S journeyman license you need to complete a Board-approved training program and one year of work experience or two years of work experience under a licensed electrician.

The exams are open book with multiple choice questions. They last three hours. Here's information on the content of the journeyman and master exams:

Subject	Percent of exam	
	Journeyman	Master
General knowledge of the electrical trade and calculations	14	16
Raceways and enclosures	11	10
Services, feeders, and branch circuits	12	12
Grounding and bonding	11	10
Conductors	11	8
Utilization and general use equipment	11	10
Special occupancies/equipment	6	6
Motor and controls	8	10
Low voltage circuits including alarms and communications	5	6
Vermont state laws and rules	11	12

The exam is given by National Assessment Institute. You can contact them at:

National Assessment Institute
2 Mount Royal Ave., Suite 250
Marlborough, MA 01752
(508) 624-0826

Electrician's license fees: It will cost you $45 for an exam. A master license costs $90 and it's $60 for either journeyman license. All licenses are good for three years.

If you have an electrician's license in New Hampshire or Massachusetts you can get a Vermont license through a reciprocity agreement Vermont has with those states. Or if you can prove you have a valid electrician's license in another state with qualifications at least equal to Vermont's, and that state accepts Vermont electricians' licenses, you can get a Vermont license without taking an exam. You'll still need to pay the fees.

Recommended Reading for the Electrical Exams

📖 *NFPA 70 - National Electrical Code*, 1996 edition, National Fire Protection Association, 1 Batterymarch Park, P. O. Box 9101, Quincy, MA 02269

📖 *Electricians' Licensing and Electrical Installation Laws and Board Rules, 26 V.S.A. Chapter 15,* Department of Labor and Industry, State Electricians' Licensing Board, Montpelier, VT 05620-3401

📖 *State of Vermont Electricians' Licensing Board Rules,* Department of Labor and Industry, Electricians' Licensing Board, Montpelier, VT 05620-3401

📖 *American Electricians' Handbook,* 12th edition, Croft, Watt, and Summers, McGraw-Hill Company

📖 *Ferm's Fast Finder Index,* Olaf G. Ferm, Ferms Finder Index Company

Type-S Journeyman Exam

📖 *NFPA 70 - National Electrical Code,* 1996 edition, National Fire Protection Association, 1 Batterymarch Park, P. O. Box 9101, Quincy, MA 02269

📖 *Electricians' Licensing and Electrical Installation Laws and Board Rules, 26 V.S.A. Chapter 15,* Department of Labor and Industry, State Electricians' Licensing Board, Montpelier, VT 05620-3401

📖 *Sign Electricians' Workbook,* 1993, James G. Stallcup, American Technical Publishers, Inc.

📖 *Design and Application of Security/Fire Alarm Systems,* 1990, John Traister, McGraw-Hill Company

📖 *NEMA Training Manual on Fire Alarm Systems,* 1992, National Electrical Manufacturers Association

📖 *NFPA 72-1993, National Fire Alarm Code,* National Fire Protection Association, 1 Batterymarch Park, P. O. Box 9101, Quincy, MA 02269

📖 *NFPA 780 - Lighting Protection Code,* 1992, National Fire Protection Association, 1 Batterymarch Park, P. O. Box 9101, Quincy, MA 02269

📖 *Fire Alarm Signaling Systems Handbook,* Bukowski and O'Laughlin, National Fire Protection Association, 1 Batterymarch Park, P. O. Box 9101, Quincy, MA 02269

You can get these books from:

Builders' Book Depot
1033 Jefferson, Suite 500
Phoenix, AZ 85034
(800) 284-3434

Plumber's Licenses

Unless you do plumbing work just on your own property or a few other special situations, you'll need a license to work in Vermont. To get information on these exceptions and an application contact:

Contact

State Plumbers Licensing Board
Department of Labor and Industry
National Life Bldg., Drawer 20
Montpelier, VT 05620-3401
(802) 828-2107

The Board issues master, journeyman, and specialists licenses. Specialist classifications are water heater specialist, heating systems specialist, and water treatment specialist.

The Board will review your application and if you're eligible they'll send you information on the exam they require for a license. For any license, you must certify that you don't owe any taxes or child support.

To qualify for the master license you need one year of experience as a journeyman or Board-approved equivalent experience. For a journeyman's license you need to complete a Board-approved apprenticeship program or Board-approved equivalent experience. For a specialist license you need to complete Board-approved instruction, training, and experience. You may be able to get the Board to accept military or out-of-state work experience for these requirements.

The exams are open book with multiple choice questions. They last three hours. Here's the content of the journeyman and master exams:

Subject	Percent of exam
Code, general knowledge	20
Installation practices, methods, and materials	15
Drainage systems, sewers	25
Water supply, backflow prevention	20
Special, indirect wastes	10
Fixtures, trim	5
Excavation	5

Fee

Plumber's license fees: It will cost you $40 for an exam. A master license costs $100, $70 for a journeyman license, and $40 for each specialist license. All licenses are good for two years.

If you have a plumber's license in New Hampshire you can get a Vermont license through a reciprocity agreement Vermont has with that state. Or if you can prove you have a valid plumber's license in another state with qualifications at least equal to Vermont's, and that state accepts Vermont plumbers' licenses, you can get a Vermont license without taking an exam. You'll still need to pay the fees.

Agency of Transportation

To bid on any Vermont Agency of Transportation construction project, you must be prequalified by the Agency. To get the Contractors' Experience Questionnaire and Financial Statement, contact:

Contact

Vermont Agency of Transportation
133 State St., Administration Bldg.
Montpelier, VT 05633-5001
(802) 828-2641

The Agency will ask you for these items:

- financial statement
- which types of work you want to be prequalified for
- your current and past work projects
- what equipment your firm owns

You'll also be asked which of the following types of work you want to be prequalified in:

Bridge construction
Bridge rehabilitation
Bridge painting
Building construction
Building demolition
Crack sealing/pavement maintenance
Foundation
Guardrail, fencing, signs

Road culverts
Shop inspection
Surfacing
Crack sealing/pavement maintenance
Surface rehabilitation
Tank/removal, replacement
Traffic signals & lighting
Water & sewer

Hazardous material removal

Landscaping

Pavement markings

Covered/timber bridge construction
  & rehabilitation

Railroads

Railroad signals

Other

The Department will use this information to give your company a prequalification rating. Prequalification is good for not more than 15 months from the date of your financial statement.

Out-of-State Corporations

Corporations doing business in Vermont must register with the Vermont Secretary of State to do business in the state. For information, contact:

Contact

Corporations
109 State Street
Montpelier, VT 05609
(802) 828-2386
Fax (802) 828-2853

Virginia

To do construction work in Virginia you need a license or certificate from the Virginia Board for Contractors. To get an application, contact:

Contact

Department of Professional and Occupational Regulation
Board for Contractors
3600 West Broad Street
P. O. Box 11066
Richmond, VA 23230-1066
(804) 367-8511
Fax: (804) 367-2474

The Board issues three types of contractor's licenses or certificates — Class A, B, or C. Usually the type of license you get will be based on the value of contracts your business will work on. Here's a description of each class:

A **Class A contractor** works on single contracts for $70,000 or more or contracts totaling $500,000 or more over a one-year period (company must have a net worth of at least $45,000).

A **Class B contractor** works on single contracts for $7,500 or more but less than $70,000 or contracts totaling $150,000 or more but less than $500,000 over a one-year period.

A **Class C contractor** works on single contracts for $1,000 or more but less than $7,500 or contracts totaling less than $150,000 over a one-year period.

The Board issues these three types of licenses in each of these general trades — building, highway/heavy, electrical, plumbing, and HVAC. It also recognizes the following specialty trades:

Alarm/security systems	Passive energy
Home improvement	Equipment/machinery
Landscape irrigation	Radon mitigation
Billboard, sign	Farm improvement
Landscape service	Recreational facility
Blast, explosive	Fire alarm systems

Commercial improvement	Refrigeration
Marine facility	Fire sprinkler
Electronic/communication service	Sewage disposal systems
Miscellaneous	Fire suppression
Elevator/escalator	Vessel construction
Modular/mobile/manufactured building	Gas fitting
Environmental monitoring well	Water well/pump

You can also get a license for asbestos and lead abatement. See below for more information on the requirements for getting a license in these trades.

To get a license you must specify a full-time employee or manager from your company to be a "qualified individual" for each trade you want to have on your license. Each person you pick must have the following experience, depending on the class of license you're applying for:

Type of license	Years of experience required
Class A license	5
Class B license	3
Class C certificate	2

To get a Class A or Class B license your company's qualified individual must pass an exam. The exam is open book with multiple choice questions. Here are the subjects on the exam and the number of questions on each subject:

Part I — Virginia section (one hour):

Subject	Number of questions
Regulation of contractors	7
Board rules, regulations	5
Building codes	5
Transaction Recovery Fund	5
Virginia erosion, sediment control regulations	2

Part II — General section (two hours):

Subject	Number of questions
Business organization	3
Estimating and bidding	10
Contract management	10
Project management	7
Risk management	3
Safety	5
Labor laws	3

Subject	Number of questions
Financial management	5
Federal, state taxes	3
Lien laws	3

Part III — Advanced section (two hours):

Subject	Number of questions
Estimating, bidding	1
Contract management	4
Project management	1
Risk management	2
Safety	2
Labor laws	4
Financial management	5
Federal, state taxes	4
Lien laws	1

For a Class B license you need to pass Parts I and II. For a Class A license you have to pass all three parts.

For electrical, plumbing, HVAC and gas fitting, the qualified individual of the company also has to get a tradesman license. At the end of this section you'll find information on how to get this license.

All the exams are administered by National Assessment Institute (NAI). They sell a reference manual you can use at the exam for $35. The manual has sections on business planning and organization, licensing, payroll, project management, contracts, referenced laws and regulations. You can contact NAI at:

National Assessment Institute
3813 Gaskins Road
Richmond, VA 23233
(800) 356-3381 (Virginia, Maryland only) / (804) 747-3297
Fax: (804) 747-5489

Contractor's license fees: It will cost you $20 nonrefundable for each part of the Class A and B exams. A Class A license costs $160 and a Class B license is $140, both nonrefundable. A Class C certificate costs $95 nonrefundable. All are good for two years.

If you have a valid license in another state the Board may grant you a Virginia license by reciprocity. Requirements for this vary so you'll need to check with the Board.

Trade Licenses (Electrical, Plumbing, HVAC and Gas Fitting)

The Tradesman Licensing Section of the Board for Contractors issues journeyman and master licenses in these trades. You can contact them at the same address as the Board (shown on page 357) but their phone number is (804) 367-2945. You must pass an exam to get any of these licenses. To qualify for a journeyman exam, you must have one of the following:

- four years of practical trade experience and 240 hours of vocational training. You can substitute each year of experience (after the four years) for 80 hours of training, up to 200 hours
- an associate degree from a Department-approved program and two years of practical experience
- an bachelor's degree from a Department-approved program and one year of practical experience
- ten years of documented Department-approved practical experience

To qualify for a master exam you must have one of the following:

- one year of experience as a licensed journeyman
- ten years of documented Department-approved practical experience

All the tradesman exams are also administered by NAI. Each exam has two parts: a regulatory section on Virginia rules and regulations and a technical section on the code for the trade. The regulatory section lasts 30 minutes and has 10 questions:

Subject	Number of questions
Standards of practice	3
Standards of conduct	2
Renewal and reinstatement	2
Definitions	1
Qualification for certification	1
Revocation of certification	1

Here's a summary of the contents and recommended reading list for each trade exam.

Electrician — Journeyman or Master

3 hours, 60 questions; master — 4 hours, 80 questions:

Subject	Number of questions	
	Journeyman	Master
General knowledge, calculations	10	19
Grounding, bonding	7	9

	Number of questions	
Subject	Journeyman	Master
Services, feeder, branch circuits	7	9
Raceways, enclosures	8	8
Motors and controls	5	8
Utilization and general use equipment	7	8
Special occupancies and equipment	6	8
Conductors	7	7
Low voltage	3	4

Gas Fitter — Journeyman or Master

2 hours, 30 questions:

Subject	Number of questions
Natural and LP gas piping	12
Pipe sizing	8
LP tanks	5
Gas piping controls	3
Testing	2

HVAC — Journeyman or Master

3 hours, 80 questions:

Subject	Number of questions
Boilers, furnaces, heaters	16
Ventilation systems, duct work	13
Refrigeration	11
Heat piping	4
Combustion air and venting	4
Controls, low voltage	4
Fuel piping	4
Insulation	3
Commercial kitchen venting	3
Underground and expansion tanks	3
Code requirements, plans, specifications	3
Test and inspection	3
Firestopping, penetration inspection	3
Trade terms	2
Energy conservation (Article 19)	2
Clearance reduction	2

Plumber — Journeyman or Master

3 hours, 70 questions:

Subject	Number of questions
Drainage waste and vent	20
General understanding of code	15
Gas piping systems	8
Water supply systems	7
Plumbing fixtures	7
Backflow prevention	5
Interceptors and traps	4
Storm drainage systems	4

Trade license fees: NAI sells practice exams for electrical, HVAC, and plumbing exams for $30 each. It will cost you $85 to take an exam. A tradesman license will cost you $45 nonrefundable and it's good for two years.

Virginia has a reciprocity agreement for electricians and plumbers licensed in North Carolina.

Recommended Reading for the Trade Exams

Electrician's Exam

📖 *NFPA 70 - National Electrical Code,* 1996 edition, National Fire Protection Association, 1 Batterymarch Park, P. O. Box 9101, Quincy, MA 02269

📖 *American Electricians Handbook,* 1996, 13th edition, Croft/Summers, McGraw-Hill Inc., Box 543, Blacklick, OH 43004-0543

Gas Fitter's Exam

📖 *1996 International Mechanical Code*

📖 *NFPA 54 - National Fuel Gas Code,* 1996, National Fire Protection Association, 11 Tracey Ave., Avon, MA 02322

HVAC Exam

📖 *1996 International Mechanical Code*

📖 *NFPA 70 - National Electrical Code,* 1996 edition, National Fire Protection Association, 1 Batterymarch Park, P. O. Box 9101, Quincy, MA 02269

📖 *Modern Refrigeration and Air Conditioning,* 1992, Althouse, Turnquist and Bracciano, The Goodheart-Willcox Company, Inc., 123 West Taft Drive, South Holland, IL 60473

Plumber's Exam

📖 *1995 International Plumbing Code (with 1996 supplement)*

📖 *NFPA 54 - National Fuel Gas Code,* 1996, National Fire Protection Association, 11 Tracey Ave., Avon, MA 02322

Asbestos Abatement Licenses

To do asbestos abatement work in Virginia, you need a license from the Board for Asbestos and Lead. To get an application, contact:

Contact

Department of Professional and Occupational Regulation
Board for Asbestos and Lead
3600 West Broad Street
P. O. Box 11066
Richmond, VA 23230-1066
(804) 367-8595 / Fax: (804) 367-2475
http://www.state.va.us/dpor

You must be licensed as a contractor or one of these individual asbestos trades — worker, supervisor, inspector, management planner, project monitor, or project designer. As a licensed asbestos contractor you and your employees must know and follow all EPA, OSHA, and Virginia rules. You must also employ only licensed asbestos personnel.

Here's information on each individual asbestos license and its training and/or experience requirements:

Worker

Required department-approved course: Initial training

Required education and/or asbestos work experience: None

Supervisor

Required department-approved course: Initial training

Required education and/or asbestos work experience: None

Inspector

Required department-approved course: Inspector training

Required education and/or asbestos work experience: One of the following: 1. bachelor's degree in a related field and six months of work experience; 2. two years college and one year of work experience; 3. high school diploma or equivalent two years of work experience

Management planner

Required department-approved course: Management planner training

Required education and/or asbestos work experience: One of the following:
1. bachelor's degree in a related field and six months of work experience;
2. two years college and one year of work experience; 3. high school diploma or equivalent two years of work experience

Project monitor

Required department-approved course: Project monitor training

Required education and/or asbestos work experience: One of the following:
1. high school diploma or equivalent and 160 hours of work experience;
2. bachelor's degree in a related field and 160 hours of work experience

Project designer

Required department-approved course: Project designer training

Required education and/or asbestos work experience: One of the following:
1. bachelor's degree in a related field and six months of work experience;
2. two years college and one year of work experience; 3. high school diploma or equivalent two years of work experience

Asbestos abatement certificate fees: It costs $50 nonrefundable to file a contractor application and the certificate will cost another $75. It will cost you $35 nonrefundable for an individual certificate. A certificate is good for one year.

The Board has reciprocity agreements with North Carolina, West Virginia, Pennsylvania, and Delaware. For information on other states, contact the Board.

Lead Abatement Certification

To do lead abatement work in Virginia, your business must be certified by the Board for Asbestos and Lead. All employees of a lead abatement business must also be certified. To get an application, contact:

Department of Professional and Occupational Regulation
Board for Asbestos and Lead
3600 West Broad Street
P. O. Box 11066
Richmond, VA 23230-1066
(804) 367-8595 / Fax: (804) 367-2475

To get a lead abatement contractor certificate you must comply with all Virginia, EPA, and OSHA regulations on lead abatement. You must also certify that all your employees have been certified by the Board. It will cost you $50 nonrefundable to file a contractor application and the certificate will cost another $75. It's good for one year.

Any individual working in lead abatement must be certified as a worker, supervisor, inspector technician, inspector/risk assessor, or planner/project designer. Here's a table showing each classification and its training and/or experience requirements.

Worker

Required department-approved course: Initial worker training

Required education and/or lead work experience: none

Supervisor

Required department-approved course: Initial supervisor training

Required education and/or lead work experience: one year experience as a lead worker or two years experience in a related field or building trade

Inspector technician

Required department-approved course: Initial inspector training

Required education and/or lead work experience: none

Inspector/risk assessor

Required department-approved course: Initial inspector/risk assessor training

Required education and/or lead work experience: one of the following: 1. one year experience and certification as an industrial hygienist, professional engineer, registered architect, or other certification in a related field; 2. two years experience and a bachelor's degree in related field; 3. three years experience and a high school diploma or equivalent; 4. 25 lead-based paint inspections over at least a three month period as a lead inspector technician and one of the following — a. certification as an industrial hygienist, professional engineer, registered architect, b. one year experience and a bachelor's degree in related field, c. two years experience and a high school diploma or equivalent

Planner/project designer

Required department-approved course: Initial planner/project designer training

Required education and/or lead work experience: none

Lead abatement certificate fees: It will cost $35 nonrefundable for an individual certificate and it's good for one year.

Department of Transportation (DOT)

To bid on Virginia Department of Transportation projects you may have to be prequalified by the Department. To get an application for prequalification call the Prequalification Section of the Department at (804) 786-2941. They'll discuss with you what procedure you should use to prequalify, depending on the type of construction work your company does. For your information, here's the address and fax number of the Department:

Virginia Department of Transportation
1401 East Broad Street
Richmond, VA 23219
(804) 786-2941
Fax: (804) 371-7896

Some of the details the Department will ask you about are your organization, personnel, financial condition, equipment, and experience. Prequalification is good for 16 months from the date of your financial statement.

Out-of-State Corporations

Out-of-state corporations must get a Certificate of Authority to do business in Virginia from the Virginia Secretary of State. To apply for this certificate contact:

Corporations Commission
Tyler Building
P. O. Box 1197
Richmond, VA 23209
(804) 371-9733
Fax: (804) 371-0744

Washington

To do construction work in Washington you must register with the Washington Department of Labor and Industries. To get an application, contact:

Contact

Department of Labor and Industries
Contractor's Registration Section
P. O. Box 44450
Olympia, WA 98504-4450
(360) 902-5226

You can register as a general or specialty contractor. If you register as a specialty contractor you can register in only two trades. Here are the trades the Department uses:

Acoustical
Air conditioning
Appliances/equipment
Asbestos
Awnings/canopies/carports/patio coverings
Boiler/steam fitting/process piping
Cabinet and millwork
Carpentry/framing
Carpet laying
Ceramic/plastic/metal file
Commercial/industrial/refrigeration
Concrete
Demolition
Drywall
Elevator
Excavating/grading
Fencing
Fire protection system
Glazing/glass
Gunite
Gutters/downspouts

Overhead/garage doors
Painting/wall covering
Paving/striping
Plastering
Plumbing
Pressure washing
Resilient floor/countertop materials/
  plastic finish/masonite
Roofing
Sanitation systems
Seal coating
Service station equipment
Sheet metal
Siding
Signs (nonelectrical)
Steel/aluminum erectors
Steel reinforcing/bar/wire mesh
Structural pest control/repair
Swimming pools/service/repair
Tanks/tank renovating
Telecom/cable wiring

House moving
Hydraulic installation/repair
Institutional/equip./
   stationary furniture/lab tables/lockers
Insulation
Irrigation/sprinkling system
Landscaping
Lathing
Machinery
Masonry
Mobile home setup
Ornamental/metals

Venetian blinds/shades/drapes
Warm air venting/ventilation/
   evap. cooling
Water conditioning equipment
Waterproofing
Weatherstripping
Welding
Well drilling
Wood floor laying/finishing
Wood stove installation
Other

You need to post a $6,000 bond if you register as a general contractor and a $4,000 bond if you register as a specialty contractor. You must also show that you have at least $20,000 of property damage insurance and $100,000 of public liability coverage. The Department will also ask you for these numbers:

- IRS employer account number
- employment security number
- industrial insurance number
- state revenue tax number

If you don't have the last three numbers, you can apply for a Washington Unified Business Identifier (UBI). To do this, contact:

Contact

Business License Service Center
P. O. Box 9034
Olympia, WA 98507-9034
(360) 753-4401

Fee

Contractor's registration fees: It'll cost you $40 to register. Registration is good for one year.

Electrician's Licenses

To do electrical work in Washington you must be licensed by the Department of Labor and Industries. The Department issues these electrical licenses — contractor, administrator, journeyman, specialty, and trainee. To get an application for a license, contact:

Contact

Department of Labor and Industries
Electrical Section
P. O. Box 44460
Olympia, WA 98504-4460
(360) 902-5269

Electrical Contractor's License

You don't have to take an exam to get an electrical contractor's license but you must agree that all the work you do will be under the direction and supervision of a licensed electrical administrator. Also you must post a $4,000 bond. The Department issues two electrical contractor licenses:

General — includes all types of wiring

Specialty — residential, pump and irrigation, signs, domestic, limited energy, and nonresidential maintenance.

Electrical contractor's license fees: It will cost you $30 for a transfer of administrator fee and $200 for a contractor's license. The license is good for two years.

Electrical Administrator's License

The Department issues these electrical administrator licenses:

General — includes all types of electrical installations

Specialty — residential, pump and irrigation, signs, domestic, limited energy, and nonresidential maintenance.

You must pass an open book, multiple choice exam to get an administrator's license. The exam has three parts — code, administration/safety, and theory. The exam for the general or residential administrator license also has a fourth section on calculations. Here's information on the percentage of each subject on the exam for each administrator license:

Subject	General
Services, feeders, branch circuits, overcurrent protection	12
Grounding, bonding	8
Raceways, enclosures	10
Conductors	6
Motor, controls	8
Utilization, general use equipment	10
Special occupancies/equipment	6
General electrical trade knowledge, calculations	20
State laws, rules, regulations	8
State general safety, health standards	8
Limited energy	4
Electronic components	—
Power supplies	—
Sound, communications	—
Fire alarms	—
Burglar alarms	—

Subject	Residential	Sign	Pump, Irrigation
Services, feeders, branch circuits, overcurrent protection	14	12	12
Grounding, bonding	8	8	10
Raceways, enclosures	8	6	8
Conductors	8	6	8
Motor, controls	2	2	12
Utilization, general use equipment	12	18	8
Special occupancies/equipment	6	8	4
General electrical trade knowledge, calculations	20	14	14
State laws, rules, regulations	10	13	10
State general safety, health standards	10	13	10
Limited energy	2	—	—
Electronic components	—	—	—
Power supplies	—	—	—
Sound, communications	—	—	—
Fire alarms	—	—	—
Burglar alarms	—	—	—

Subject	Domestic appliance	Non-residential	Limited energy
Services, feeders, branch circuits, overcurrent protection	12	12	6
Grounding, bonding	10	8	8
Raceways, enclosures	12	8	8
Conductors	12	8	8
Motor, controls	—	10	—
Utilization, general use equipment	22	10	4
Special occupancies/equipment	—	6	2
General electrical trade knowledge, calculations	14	12	10
State laws, rules, regulations	11	12	12
State general safety, health standards	11	12	12
Limited energy	—	2	—
Electronic components	—	—	6
Power supplies	—	—	7
Sound, communications	—	—	7
Fire alarms	—	—	8
Burglar alarms	—	—	8

Electrical administrator license fees: It costs $85 to file the application for an administrator's license, $25 nonrefundable for the application and $60 for the license. It'll cost you $60 for the exam. A license is good for two years but it expires on your birthday. Washington doesn't have any reciprocity agreement with any other state for this type of license.

Journeyman and Specialty Electrician's Licenses

The Department issues journeyman and these specialty licenses — residential, sign, pump and irrigation, domestic appliance, nonresidential maintenance, and limited energy.

To apply for the journeyman's licensing exam you need to have four years (7,200 hours) of experience including at least two years (3,600 hours) in commercial or industrial electrical installation work. To apply for the specialty licensing exam you need 3,600 hours in the specialty.

You must pass an open book, multiple choice exam for any of these licenses. The journeyman exam lasts three hours and a specialty exam lasts two hours. Here's the percentage of each subject on the exam for each administrator license:

Subject	Journeyman
Grounding, bonding	10
Services, feeders, branch circuits, overcurrent protection	12
Raceways, enclosures	12
Conductors	12
Motor, controls	10
Utilization, general use equipment	10
Special occupancies/equipment	6
General electrical trade knowledge, calculations	18
Limited energy	4
State laws, rules, regulations	6
Electronic components	—
Power supplies	—
Sound, communications	—
Fire alarms	—
Burglar alarms	—

Subject	Residential	Sign	Pump, Irrigation
Grounding, bonding	10	10	10
Services, feeders, branch circuits, overcurrent protection	16	14	14
Raceways, enclosures	8	12	12
Conductors	14	14	12
Motor, controls	4	10	18
Utilization, general use equipment	14	14	12
Special occupancies/equipment	6	4	4
General electrical trade knowledge, calculations	22	18	14
Limited energy	2	—	—
State laws, rules, regulations	4	4	4
Electronic components	—	—	—
Power supplies	—	—	—
Sound, communications	—	—	—
Fire alarms	—	—	—
Burglar alarms	—	—	—

Subject	Domestic appliance	Non-residential	Limited energy
Grounding, bonding	10	10	8
Services, feeders, branch circuits, overcurrent protection	10	10	4
Raceways, enclosures	12	10	8
Conductors	16	12	8
Motor, controls	6	12	—
Utilization, general use equipment	22	12	6
Special occupancies/equipment	—	8	2
General electrical trade knowledge, calculations	20	19	16
Limited energy	—	4	—
State laws, rules, regulations	4	4	4
Electronic components	—	—	8
Power supplies	—	—	6
Sound, communications	—	—	10
Fire alarms	—	—	10
Burglar alarms	—	—	10

The reference list for the journeyman and specialty exam is the same as the one for the administrator exam.

Fee

Journeyman and specialty electrician's license fees: It will cost you $65 to file the application for an administrator's license, $25 nonrefundable for the application and $40 for the license. It costs $45 for the exam. A license is good for up to three years but the period will depend on when you get the license. Washington doesn't have any reciprocity agreement with any other state for this type of license.

If you've had a journeyman license from Alaska, Idaho, South Dakota, North Dakota, Utah, Wyoming, Oregon, or Massachusetts for at least a year and took an exam to get it, the Department may recognize your license.

Electrical Trainee's Certificate

You must get a training certificate to learn the electrical construction trade in Washington. The certificate, which is good for one year, will cost you $20.

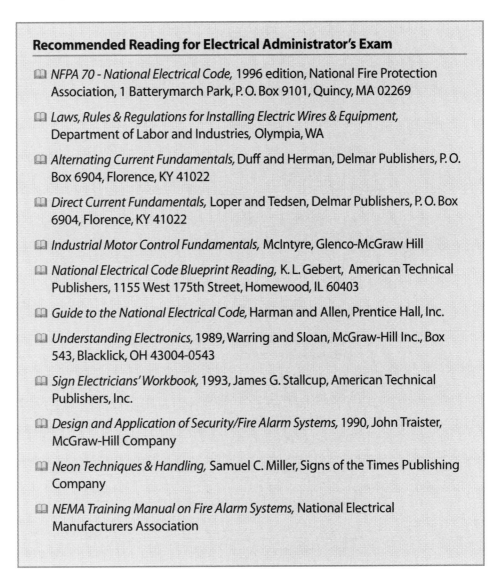

Recommended Reading for Electrical Administrator's Exam

- *NFPA 70 - National Electrical Code,* 1996 edition, National Fire Protection Association, 1 Batterymarch Park, P. O. Box 9101, Quincy, MA 02269

- *Laws, Rules & Regulations for Installing Electric Wires & Equipment,* Department of Labor and Industries, Olympia, WA

- *Alternating Current Fundamentals,* Duff and Herman, Delmar Publishers, P. O. Box 6904, Florence, KY 41022

- *Direct Current Fundamentals,* Loper and Tedsen, Delmar Publishers, P. O. Box 6904, Florence, KY 41022

- *Industrial Motor Control Fundamentals,* McIntyre, Glenco-McGraw Hill

- *National Electrical Code Blueprint Reading,* K. L. Gebert, American Technical Publishers, 1155 West 175th Street, Homewood, IL 60403

- *Guide to the National Electrical Code,* Harman and Allen, Prentice Hall, Inc.

- *Understanding Electronics,* 1989, Warring and Sloan, McGraw-Hill Inc., Box 543, Blacklick, OH 43004-0543

- *Sign Electricians' Workbook,* 1993, James G. Stallcup, American Technical Publishers, Inc.

- *Design and Application of Security/Fire Alarm Systems,* 1990, John Traister, McGraw-Hill Company

- *Neon Techniques & Handling,* Samuel C. Miller, Signs of the Times Publishing Company

- *NEMA Training Manual on Fire Alarm Systems,* National Electrical Manufacturers Association

---

Recommended Reading for Electrical Administrator's Exam (continued)

📖 *Master Electrician Home Study Guide,* Barclay, Accent Printing

📖 *Ferm's Fast Finder Index,* Olaf G. Ferm, Ferms Finder Index Company

📖 *Industrial Motor Control Fundamentals,* Herman and Alerich, Delmar Publishers, P. O. Box 6904, Florence, KY 41022

📖 *American Electricians' Handbook,* Croft, Watt, and Summers, McGraw-Hill Company

📖 *Fire Alarm Signaling Systems Handbook,* Bukowski and O'Laughlin, National Fire Protection Association

---

Department of Transportation (DOT)

To bid on any Washington Department of Transportation construction project, you must be prequalified by the Department. To get their Standard Questionnaire and Financial Statement, contact:

Contact

Prequalification Officer
Department of Transportation
P. O. Box 47360
Olympia, WA 98504-7360
(360) 705-7837

The Transportation Department will ask you for these items:

- current financial statement
- which types of work you want to be prequalified for
- your current and past work projects
- what equipment your firm owns

You'll also be asked which of the following types of work you want to be prequalified in:

Clearing, grubbing, grading, draining	Tunnels and shaft excavation
Production and placing crushed materials	Piledriving
Bituminous surface treatment	Concrete surface treatment
Asphalt concrete paving	Fencing
Cement concrete paving	Bridge deck repair
Bridges and structures	Deck seal
Buildings	Signing
Painting	Sign structures and signs
Traffic signals	Slurry diaphragm and cut-off walls

Structural tile cleaning

Guardrails

Pavement marking

Demolition

Drilling and blasting

Sewers and water mains

Illumination and general electric

Cement concrete curb and gutter

Asphalt concrete curb and gutter

Riprap and rock walls

Concrete structures except bridges

Surveying

Water distribution and irrigation

Landscaping

Engineering

Erosion control

Precast median barrier

Permanent tie-back anchor

Impact attenuators

Paint striping

Wire mesh slope protection

Gabion and gabion construction

Electronics

Mechanical

Asbestos abatement

Hazardous waste material

Concrete restoration

Concrete sawing, coring and grooving

Dredging

Marine work

Ground modification

Well drilling

Sewage disposal

Traffic control

Railroad construction

Steel fabrication

Street cleaning

Materials transporting

Sandblasting and steam cleaning

The Department will use this information to give your company a prequalification rating. Prequalification is good for one year.

Out-of-State Corporations

Corporations doing business in Washington must register with the Washington Secretary of State. For information, contact:

Contact

Corporations Division
Office of the Secretary of State
505 East Union
Olympia, WA 98504
(360) 753-7115

West Virginia

You must be licensed to work on construction in West Virginia. To get an application packet, contact:

Contact

West Virginia Contractor Licensing Board
Building 3, Room 319
State Capitol Complex
Charleston, WV 25305
(304) 558-7890

You'll need to have unemployment and workers' compensation coverage in your home state or West Virginia. You also have to post a wage bond unless you work only in residential contract work or have been doing construction work in West Virginia for at least five years before you apply for a license. The bond must equal the total of your gross payroll for four weeks at full capacity or production plus 15% for benefits.

It's possible to get the West Virginia Contractor Licensing Board to accept your license in another state depending on the qualifications you passed to be licensed there. If so, you won't have to take any exam but you will have to pay West Virginia's licensing fees.

Construction Contractor's Licenses

The Licensing Board uses these contractor classifications:

Electrical   HVAC
Piping   Multifamily
General building   Plumbing
General engineering   Residential building

The Board specialty classifications are:

Drywall   Siding
Excavation   Roofing
Landscaping   Low voltage systems

Masonry

Remodeling and repair

Asphalt

Concrete

Residential pools

Manufactured home installation

Steel erection

Sprinklers and fire protection

Others as determined by the Board

Fee

Contractor licence fees: A license is good for one year and costs $90. But that doesn't include fees for the exams the Board requires for a license.

You have to pass a business and law exam and a trade exam to get a license. The exams are given by Block & Associates. You can contact them at:

Contact

Block & Associates

331 12th Street

Dunbar, WV 25064

(304) 768-2290

Fax (304) 768-2494

The business and law exam is open book with multiple choice questions on state and federal laws and business, financial and accounting practices. Here are the subjects on the business and law exam and the approximate percentage of each subject on the exam:

Subject	Percent of exam
Project management	20
Contract management	20
License law	10
Financial management	10
Safety regulations	6
Labor laws	10
Risk management	8
Tax law	8
Lien law	4
Business planning and organization	4

Fee

Contractor exam fees: Block & Associates has prepared the West Virginia Contractors Reference Manual that you can use at the exam. They sell it for $40 plus tax and shipping. The exam will cost you $40 plus tax. They also sell a practice business and law exam for $28 plus tax.

You'll also have to pass an exam on the trade you want a license in. A trade exam costs $40 plus tax. All the trade exams are open book. You can use published and bound reference materials at the exam but not loose papers, notes, or exam study guides.

Here's information on all of the trade exams. You can get the references listed below from:

Contact

Contractors Book Service
P.O. Box 1008
Dunbar, WV 25064
(800) 408-5745

Builders' Book Depot
1033 Jefferson, Suite 500
Phoenix, AZ 85034
(800) 284-3434

Professional Booksellers
2200 21st Avenue
South Nashville, TN 3721
(800) 572-8878

General Building Contractor

Subject	Percent of exam
Carpentry	28
Concrete	20
Site work	12
Structural steel and rebar	14
Masonry	10
Roofing	6
Drywall	6
Insulation	4

General Engineering

Subject	Percent of exam
Excavating, grading	25
Concrete construction	20
Structural steel and rebar	15
Utilities	15
Asphalt paving	15
Drilling, demolition, pools, etc.	10

HVAC Contractor

Subject	Percent of exam
Warm-air heating	15
Air conditioning	15
Service	10
Controls	10
Sizing and estimating	10
Low pressure boilers	7
Refrigeration	6
Ducts	6
Chimneys, flues, vents	8
Gas piping	8
Ventilation	5

Multifamily Building Contractor

Subject	Percent of exam
Rough carpentry	20
Finish carpentry	19
Concrete and rebar	17
Excavation and site work	14
Roofing	10
Masonry	10
Electrical, HVAC, plumbing	10

Piping Contractor

Subject	Percent of exam
Fuel and gas piping	20
Hazardous chemicals, process piping	20
Water and steam piping	20
Excavation and grading	10
Sizing and estimating	9
Materials	10
Welding	10
Safety	1

Plumbing Contractor

Subject	Percent of exam
Drains, vents/savers	17
Water supply, backflow prevention	17
Code knowledge	13
Installation practices, methods, materials	10
Septic tanks special wastes, roof drains	10
Inspection, testing	8
Natural gas piping	9
Fixtures and trim	5
Excavation	7
Estimating	4

Residential Building Contractor

Subject	Percent of exam
Rough carpentry	24
Concrete and rebar	20
Electrical, plumbing, HVAC	5
Excavation and site work	10
Finish carpentry	15
Roofing	12
Masonry	9
Lightweight metal framing	5

Drywall Contractor

Subject	Percent of exam
Installation	37
Taping and texturing	20
Materials	18
Metal studs	15
Special applications	5
Estimating, plans	5

Excavation Contractor

Subject	Percent of exam
Equipment, techniques	50
Compaction	18
Soil types	10
Estimating	15
Blasting	7

Landscaping Contractor

Subject	Percent of exam
Sprinklers, irrigation	26
Grading, drainage	14
Soils and amendments	10
Trees, shrubs	22
Landscaping - grass	16
Electrical	4
Fencing, carpentry	4
Flatwork	4

Low Voltage Systems Contractor

Subject	Percent of exam
National Electrical Code	38
Equipment	28
General electrical knowledge	16
Systems design	10
Signals transmission, conductors	8

Manufactured Home Installation Contractor

Subject	Percent of exam
Foundation installation	25
Ground anchors	20
Utilities	17
Connection of multi-wide homes	15
Site preparation	10
Exterior finish	8
Federal requirements	5

Masonry Contractor

Subject	Percent of exam
Unit masonry	36
Mortar, grout	28
Tools	6
Restoration, cleaning	6
Safety	8
Stone masonry	2
Estimating, plans, specifications	8
Accessories	6

Remodeling and Repair Contractor

Subject	Percent of exam
Masonry	10
Associate trades	14
Rough carpentry	32
Concrete	14
Finish carpentry	20
Roofing	10

Residential Asphalt Contractor

Subject	Percent of exam
Base preparation	25
Placing techniques	25
Excavation, grading	20
Finishing and coatings	18
Properties of asphalt	12

Residential Concrete Contractor

Subject	Percent of exam
Layout, excavation	14
Formwork	18
Pour, place, cure	26
Estimating, plan, specifications	12
Batching	14
Reinforcing steel	16

Residential Pools Contractor

Subject	Percent of exam
Concrete	28
Site preparation, excavation	10
Water treatment, circulation	15
Tiling	10
Fuel gas piping, heaters	10
Electrical systems	10
Backflow prevention	7
Chlorine piping	5
Drainage, groundwater control	3
Fencing	2

Residential Siding Contractor

Subject	Percent of exam
Installation methods	40
Wood	30
Metal	20
Vinyl	10

Roofing Contractor

Subject	Percent of exam
Composition shingles	26
Built-up roofing	26
Mineral, clay shingles	12
Waterproofing	12
Metal roofing, flashing	10
Wood shingles, shakes	10
Insulation	4

Steel Erection Contractor

Subject	Percent of exam
Steel erection	30
Structural steel materials	20
Hoisting, rigging	20
Welding, cutting	15
Tools, equipment	15

Sprinklers and Fire Protection Contractor

Subject	Percent of exam
Sprinkler systems	45
Wet, dry chemical systems	25
Water supply, storage, distribution	10
General code, trade knowledge	10
Flammable storage	5
Commercial kitchen equipment	5

Electrical Contractor

Subject	Percent of exam
General knowledge and calculations	22
Grounding and bonding	11
Services, feeders, branch circuits	13
Raceways, enclosures	12
Motors and controls	11
Utilization and general use equipment	11
Special occupancies and equipment	7
Conductors	10
Low voltage circuits	3

You will also need an individual electrician's license to do electrical work in West Virginia. To get an application, contact:

Contact

State Fire Marshal
Electrician's License Section
1207 Quarrier Street, 2nd Floor
Charleston, WV 25301
(304) 558-2191

The Section issues master, journeyman, and specialty electrician's licenses. You have to pass an exam to get a license. To qualify for the master exam you need five years in residential, commercial, and industrial electrical work. To qualify for the journeyman exam you need to complete one of the following:

- four years of work experience under a master electrician

- an approved apprenticeship and training program

- 1,080 hours of approved vocational education

To qualify for a specialty exam you need at least two years of experience in the specialty. You can apply one year of approved education to the work requirement.

The exam is open book on the *National Electrical Code* and lasts four hours. Here's a summary that shows the number of questions on each exam:

License	Questions on exam
Master	40 multiple choice, 12 calculations
Journeyman	75 multiple choice, 5 calculations
HVAC	40 multiple choice
Electric sign	40 multiple choice
Single family residence	79 multiple choice
Low voltage	67 multiple choice
Fiber optics	40 multiple choice

Fee

An exam will cost you $25. A license costs $50 and it's good for one year. All licenses expire on June 30.

The Section doesn't have any reciprocity agreements with other states for these licenses.

Recommended Reading for the Contractor's Trade Exam

General Building Contractor's Exam

- *BOCA National Building Code,* Building Officials and Code Administrators Int. Inc., 405 West Flossmoor Road, Country Club Hills, IL, 60478-4900

- *Carpentry and Building Construction,* 1993, Feirer, Hutchings & Feirer, McGraw-Hill Inc., Box 543, Blacklick, OH 43004-0543

- *Design and Control of Concrete Mixtures,* 1988/1990, 13th edition, Portland Cement Association, 5420 Old Orchard Road, Skokie, IL 60077-1083

- *Excavation & Grading Handbook,* 1991, Nicholas E. Capachi, Craftsman Book Company, 6058 Corte del Cedro, P.O. Box 6500, Carlsbad, CA 92018

- *Manual of Steel Construction,* 1989, 9th edition, American Institute of Steel Construction, P. O. Box 806276, Chicago, IL 60680-4124

- *Modern Carpentry,* 1992, Willis H. Wagner, The Goodheart-Willcox Company, Inc.

- *Modern Masonry,* 1991, C. E. Kicklighter, The Goodheart-Willcox Company, Inc.

- *NRCA Roofing and Waterproofing Manual,* 1989, 3rd edition, National Roofing Contractor's Association, 10255 W. Higgins, Rd., Suite #600, Rosemont, IL 60018-5607

Recommended Reading for the Contractor's Trade Exam (continued)

General Building Contractor's Exam (continued)

📖 *Code of Federal Regulations - Title 29, Part 1926 (OSHA),* July 1995, Superintendent of Documents, U.S. Government Printing Office, Washington, DC 20402

📖 *Placing Reinforcing Bars, Recommended Practices,* 1992, 6th edition, Concrete Reinforcing Steel Institute, P. O. Box 6996, Alpharetta, GA 30239-6996

General Engineering Exam

📖 *MS-8- Asphalt Paving Manual,* 3rd edition, Asphalt Institute, P. O. Box 14052, Lexington, KY 40512-4052

📖 *Building Construction Materials and Types of Construction,* 1981, W. C. Huntington, R. E. Mackadeit, John Wiley and Sons

📖 *Construction of Structural Steel Building Frames ,* W. G. Rapp

📖 *Design and Control of Concrete Mixtures,* 1988/1990, 13th edition, Portland Cement Association, 5420 Old Orchard Road, Skokie, IL 60077-1083

📖 *Excavation & Grading Handbook,* 1991, Nicholas E. Capachi, Craftsman Book Company, 6058 Corte del Cedro, P.O. Box 6500, Carlsbad, CA 92018

📖 *Explosives and Rock Blasting,* 1987, Atlas Powder Company

📖 *Highway Materials, Soils and Concrete,* 1983, 2nd edition, Nadon, H. N. Atkins, Reston Publishing Co.

📖 *Moving the Earth,* 3rd edition, 1987, H. L. Nichols, Jr., North Castle Books

📖 *Code of Federal Regulations - Title 29, Part 1926 (OSHA),* July 1995, Superintendent of Documents, U.S. Government Printing Office, Washington, DC 20402

📖 *Placing Reinforcing Bars, Recommended Practices,* 1992, 6th edition, Concrete Reinforcing Steel Institute, P. O. Box 6996, Alpharetta, GA 30239-6996

HVAC Contractor's Exam

📖 *Heating, Ventilation, and Air Conditioning Library, Vol. I, II, and III,* Theodore Audel & Co., Division of The Bobbs-Merrill Company Inc.

📖 *Modern Refrigeration and Air Conditioning,* Althouse, Turnquist and Bracciano, The Goodheart-Willcox Company, Inc.

📖 *Code of Federal Regulations - Title 29, Part 1926 (OSHA),* July 1995, Superintendent of Documents, U.S. Government Printing Office, Washington, DC 20402

Recommended Reading for the Contractor's Trade Exam (continued)

HVAC Contractor's Exam (continued)

📖 *Trane Air Conditioning Manual,* 1965, Trane Company, 8929 Western Way, Suite #1, Jacksonville, FL 32256

Multifamily Building Contractor's Exam

📖 *BOCA National Building Code,* Building Officials and Code Administrators Int. Inc., 405 West Flossmoor Road, Country Club Hills, IL, 60478-4900

📖 *Carpentry and Building Construction,* 1993, Feirer, Hutchings & Feirer, McGraw-Hill Inc., Box 543, Blacklick, OH 43004-0543

📖 *Concrete Form Construction,* 1977, Moore, Delmar Publishers, P. O. Box 6904, Florence, KY 41022

📖 *Concrete Masonry Handbook for Architects, Engineers, Builders,* 1985, Randal & Panarese, Portland Cement Association, 5420 Old Orchard Road, Skokie, IL 60077-1083

📖 *Design and Control of Concrete Mixtures,* 1988/1990, 13th edition, Portland Cement Association, 5420 Old Orchard Road, Skokie, IL 60077-1083

📖 *Excavation & Grading Handbook,* 1991, Nicholas E. Capachi, Craftsman Book Company, 6058 Corte del Cedro, P.O. Box 6500, Carlsbad, CA 92018

📖 *Modern Carpentry,* 1992, Willis H. Wagner, The Goodheart-Willcox Company, Inc.

📖 *Modern Masonry,* 1991, C. E. Kicklighter, The Goodheart-Willcox Company, Inc.

📖 *NFPA 70 - National Electrical Code,* 1996 edition, National Fire Protection Association, 1 Batterymarch Park, P. O. Box 9101, Quincy, MA 02269

📖 *NRCA Roofing and Waterproofing Manual,* 1989, 3rd edition, National Roofing Contractor's Association, 10255 W. Higgins, Rd., Suite #600, Rosemont, IL 60018-5607

📖 *Roofers Handbook,* 1986, William Johnson, Craftsman Book Company, 6058 Corte del Cedro, P.O. Box 6500, Carlsbad, CA 92018

Piping Contractor's Exam

📖 *BOCA National Mechanical Code,* Building Officials and Code Administrators Int. Inc., 405 West Flossmoor Road, Country Club Hills, IL, 60478-4900

📖 *Code of Federal Regulations - Title 29, Part 1926 (OSHA),* July 1995, Superintendent of Documents, U.S. Government Printing Office, Washington, DC 20402

Recommended Reading for the Contractor's Trade Exam (continued)

Piping Contractor's Exam (continued)

📖 *Excavation & Grading Handbook,* 1991, Nicholas E. Capachi, Craftsman Book Company, 6058 Corte del Cedro, Box 6500, Carlsbad, CA 92018

📖 *Modern Welding,* McGraw-Hill Inc., P.O. Box 543, Blacklick, OH 43004-0543

Plumbing Contractor's Exam

📖 *BOCA National Mechanical Code,* Building Officials and Code Administrators Int. Inc., 405 West Flossmoor Road, Country Club Hills, IL, 60478-4900

📖 *BOCA National Plumbing Code,* 1990 edition, Building Officials and Code Administrators Int. Inc., 405 West Flossmoor Road, Country Club Hills, IL, 60478-4900

📖 *The Plumbers Handbook,* 1985 G. K. Hall & Co.

Residential Building Contractor's Exam

📖 *BOCA National Building Code,* Building Officials and Code Administrators Int. Inc., 405 West Flossmoor Road, Country Club Hills, IL, 60478-4900

📖 *Carpentry and Building Construction,* 1993, Feirer, Hutchings & Feirer, McGraw-Hill Inc., Box 543, Blacklick, OH 43004-0543

📖 *Design and Control of Concrete Mixtures,* 1988/1990, 13th edition, Portland Cement Association, 5420 Old Orchard Road, Skokie, IL 60077-1083

📖 *Excavation & Grading Handbook,* 1991, Nicholas E. Capachi, Craftsman Book Company, 6058 Corte del Cedro, P.O. Box 6500, Carlsbad, CA 92018

📖 *Gypsum Construction Handbook,* 1992, United States Gypsum Company

📖 *Modern Carpentry,* 1992, Willis H. Wagner, The Goodheart-Willcox Company, Inc.

📖 *Modern Masonry,* 1991, C. E. Kicklighter, The Goodheart-Willcox Company, Inc.

📖 *NRCA Roofing and Waterproofing Manual,* 1989, 3rd edition, National Roofing Contractor's Association, 10255 W. Higgins, Rd., Suite #600, Rosemont, IL 60018-5607

📖 *Code of Federal Regulations - Title 29, Part 1926 (OSHA),* July 1995, Superintendent of Documents, U.S. Government Printing Office, Washington, DC 20402

📖 *Placing Reinforcing Bars, Recommended Practices,* 1992, 6th edition, Concrete Reinforcing Steel Institute, P. O. Box 6996, Alpharetta, GA 30239-6996

Recommended Reading for the Contractor's Trade Exam (continued)

Drywall Contractor's Exam

📖 *Gypsum Construction Handbook,* 1992, United States Gypsum Company

📖 *Using Gypsum Board for Walls and Ceilings,* 1991, 1993, Gypsum Association, 810 First Street NE, Washington, DC 20002

📖 *Code of Federal Regulations - Title 29, Part 1926 (OSHA),* July 1995, Superintendent of Documents, U.S. Government Printing Office, Washington, DC 20402

Excavation Contractor's Exam

📖 *BOCA National Building Code,* Building Officials and Code Administrators Int. Inc., 405 West Flossmoor Road, Country Club Hills, IL, 60478-4900

📖 *Carpentry and Building Construction,* 1993, Feirer, Hutchings & Feirer, McGraw-Hill Inc., Box 543, Blacklick, OH 43004-0543

📖 *Explosives and Rock Blasting,* 1987, Atlas Powder Company

📖 *Excavation and Grading Handbook,* 1991, Nicholas E. Capachi, Craftsman Book Company, 6058 Corte del Cedro, Box 6500, Carlsbad, CA 92018

📖 *Code of Federal Regulations - Title 29, Part 1926 (OSHA),* July 1995, Superintendent of Documents, U.S. Government Printing Office, Washington, DC 20402

📖 *Principles and Practices of Heavy Construction,* 4th edition, R. C. Smith and C. K. Andres, Prentice-Hall Publishers, 200 Old Tappan, Old Tappan, NJ -07675

📖 *Moving the Earth,* 3rd edition, 1987, H. L. Nichols, Jr., North Castle Books

Landscaping Contractor's Exam

📖 *Building Fences of Wood, Stone, Metal, and Plants,* 1987, John Williamson Publishing

📖 *Landscape Operations,* 1993, L. G. Hannebaum, Reston Publishing Co.

📖 *Landscaping Principles & Practices,* 1992, Ingles

📖 *Turf Irrigation Manual,* 1994, Choate

Low Voltage Systems Contractor's Exam

📖 *American Electricians Handbook,* 1996, 13th edition, Croft/Summers, McGraw-Hill Inc., Box 543, Blacklick, OH 43004-0543

Recommended Reading for the Contractor's Trade Exam (continued)

Low Voltage Systems Contractor's Exam (continued)

📖 *Design and Application of Security/Fire Alarm Systems,* John Traister, McGraw-Hill Company

📖 *Direct Current Fundamentals,* Loper and Tedsen, Delmar Publishers, P. O. Box 6904, Florence, KY 41022

📖 *Fire Alarm Signaling Systems Handbook,* Bukowski and O'Laughlin, National Fire Protection Association, 1 Batterymarch Park, P. O. Box 9101, Quincy, MA 02269

📖 *NFPA 70 - National Electrical Code,* 1996 edition, National Fire Protection Association, 1 Batterymarch Park, P. O. Box 9101, Quincy, MA 02269

📖 *NFPA 72-1993, National Fire Alarm Code,* National Fire Protection Association, 1 Batterymarch Park, P. O. Box 9101, Quincy, MA 02269

📖 *NFPA 72 - Installation, Maintenance, and Use of Protective Signaling Systems,* National Fire Protection Association, 1 Batterymarch Park, P. O. Box 9101, Quincy, MA 02269

📖 *NFPA 72E - Automatic Fire Detectors,* National Fire Protection Association, 1 Batterymarch Park, P. O. Box 9101, Quincy, MA 02269

📖 *NFPA 74 - Household Fire Warning Equipment,* National Fire Protection Association, 1 Batterymarch Park, P. O. Box 9101, Quincy, MA 02269

Manufactured Home Installation Contractor's Exam

📖 *NCSBCS Standard for Manufactured Home Installation, A225.1,* National Conference on States on Building Codes and Standards

📖 *Manufactured Home Installation Study Guide,* 1995, National Conference on States on Building Codes and Standards

Masonry Contractor's Exam

📖 *BOCA National Building Code, Building Officials and Code Administrators Int. Inc.,* 405 West Flossmoor Road, Country Club Hills, IL, 60478-4900

📖 *Concrete Masonry Handbook,* 1991, Randall & Panarese, Portland Cement Association, 5420 Old Orchard Road, Skokie, IL 60077-1083

📖 *Masonry Skills,* 1990, R. T. Kreh, Sr.

📖 *Modern Masonry,* 1991, C. E. Kicklighter, The Goodheart-Willcox Company, Inc.

Recommended Reading for the Contractor's Trade Exam (continued)

Masonry Contractor's Exam (continued)

📖 *Code of Federal Regulations - Title 29, Part 1926 (OSHA)*, July 1995, Superintendent of Documents, U.S. Government Printing Office, Washington, DC 20402

Remodeling and Repair Contractor's Exam

📖 *BOCA National Building Code,* Building Officials and Code Administrators Int. Inc., 405 West Flossmoor Road, Country Club Hills, IL, 60478-4900

📖 *Carpentry and Building Construction,* 1993, Feirer, Hutchings & Feirer, McGraw-Hill Inc., Box 543, Blacklick, OH 43004-0543

📖 *Design and Control of Concrete Mixtures,* 1988/1990, 13th edition, Portland Cement Association, 5420 Old Orchard Road, Skokie, IL 60077-1083

📖 *Modern Masonry,* 1991, C. E. Kicklighter, The Goodheart-Willcox Company, Inc.

📖 *Modern Carpentry,* 1992, Willis H. Wagner, The Goodheart-Willcox Company, Inc.

📖 *Gypsum Construction Handbook,* 1992, United States Gypsum Company

📖 *NRCA Roofing and Waterproofing Manual,* 1989, 3rd edition, National Roofing Contractor's Association, 10255 W. Higgins, Rd., Suite #600, Rosemont, IL 60018-5607

📖 *Code of Federal Regulations - Title 29, Part 1926 (OSHA),* July 1995, Superintendent of Documents, U.S. Government Printing Office, Washington, DC 20402

Residential Asphalt Contractor's Exam

📖 *MS-8- Asphalt Paving Manual,* 3rd edition, Asphalt Institute, P. O. Box 14052, Lexington, KY 40512-4052

📖 *Carpentry and Building Construction,* 1993, Feirer, Hutchings & Feirer, McGraw-Hill Inc., Box 543, Blacklick, OH 43004-0543

📖 *Excavation & Grading Handbook,* 1991, Nicholas E. Capachi, Craftsman Book Company, 6058 Corte del Cedro, P.O. Box 6500, Carlsbad, CA 92018

📖 *Highway Materials, Soils and Concrete,* 1983, 2nd edition, Nadon, H. N. Atkins, Reston Publishing Co.

📖 *Moving the Earth,* 3rd edition, 1987, H. L. Nichols, Jr., North Castle Books

📖 *Methods and Materials of Residential Construction,* 1985, James E. Russell, Prentice-Hall, Inc.

Recommended Reading for the Contractor's Trade Exam (continued)

Residential Concrete Contractor's Exam (continued)

- *BOCA National Building Code,* Building Officials and Code Administrators Int. Inc., 405 West Flossmoor Road, Country Club Hills, IL, 60478-4900

- *Concrete Construction & Estimating,* 1991, Craig Avery, Craftsman Book Company, 6058 Corte del Cedro, P.O. Box 6500, Carlsbad, CA 92018

- *Concrete Form Construction,* 1977, Moore, Delmar Publishers, P. O. Box 6904, Florence, KY 41022

- *Design and Control of Concrete Mixtures,* 1988/1990, 13th edition, Portland Cement Association, 5420 Old Orchard Road, Skokie, IL 60077-1083

- *Code of Standard Practice for Steel Buildings and Bridges,* 1992, American Institute of Steel Construction, P. O. Box 806276, Publication Department, Chicago, IL 60680-4124

- *Placing Reinforcing Bars, Recommended Practices,* 1992, 6th edition, Concrete Reinforcing Steel Institute, P. O. Box 6996, Alpharetta, GA 30239-6996

Residential Pools Contractor's Exam

- *BOCA National Mechanical Code,* Building Officials and Code Administrators Int. Inc., 405 West Flossmoor Road, Country Club Hills, IL, 60478-4900

- *BOCA National Plumbing Code,* Building Officials and Code Administrators Int. Inc., 405 West Flossmoor Road, Country Club Hills, IL, 60478-4900

- *Design and Control of Concrete Mixtures,* 1988/1990, 13th edition, Portland Cement Association, 5420 Old Orchard Road, Skokie, IL 60077-1083

- *Excavation & Grading Handbook,* 1991, Nicholas E. Capachi, Craftsman Book Company, 6058 Corte del Cedro, P.O. Box 6500, Carlsbad, CA 92018

- *Guide to Shotcrete, ACI 506R-90,* 1990, American Concrete Institute, P. O. Box 19150, Detroit, MI 48219-0150

- *NFPA 70 - National Electrical Code,* 1996 edition, National Fire Protection Association, 1 Batterymarch Park, P. O. Box 9101, Quincy, MA 02269

- *NSF Standard No. 50*

- *NSPI Pool/Spa Operators Handbook,* 1983-90, National Swimming Pool Foundation

- *Standard for Residential Swimming Pools,* 1987, National Spa and Pool Institute

- *Placing Reinforcing Bars, Recommended Practices,* 1992, 6th edition, Concrete Reinforcing Steel Institute, P. O. Box 6996, Alpharetta, GA 30239-6996

Recommended Reading for the Contractor's Trade Exam (continued)

Residential Siding Contractor's Exam

📖 *Aluminum Siding Application Manual,* 1987, Aluminum Association of America

📖 *Carpentry and Building Construction,* 1993, Feirer, Hutchings & Feirer, McGraw-Hill Inc., Box 543, Blacklick, OH 43004-0543

📖 *Code of Standard Practice for Steel Buildings and Bridges,* 1992, American Institute of Steel Construction, P. O. Box 806276, Publication Department, Chicago, IL 60680-4124

📖 *Plastering Skills,* 1984, Branden/Hartsell, American Technical Publishes, Inc.

Roofing Contractor's Exam

📖 *NRCA Roofing and Waterproofing Manual,* 1989, 3rd edition, National Roofing Contractor's Association, 10255 W. Higgins, Rd., Suite #600, Rosemont, IL 60018-5607

📖 *Roofers Handbook,* 1986, William Johnson, Craftsman Book Company, 6058 Corte del Cedro, P.O. Box 6500, Carlsbad, CA 92018

Steel Erection Contractor's Exam

📖 *Construction of Structural Steel Building Frames ,* W. G. Rapp

📖 *Manual of Steel Construction,* 1989, 9th edition, American Institute of Steel Construction, P. O. Box 806276, Chicago, IL 60680-4124

📖 *Materials of Construction,* R. C. Smith, 1979

📖 *Principles and Practices of Heavy Construction,* 4th edition, R. C. Smith and C. K. Andres, Prentice-Hall Publishers, 200 Old Tappan, Old Tappan, NJ -07675

📖 *Rigging for Commercial Construction,* 1983, Resource Systems International, Reston Publishing, Inc.

📖 *Welding Fundamentals and Procedures,* 1984, Gaylen, John Wiley & Sons, Inc.

Sprinklers and Fire Protection Contractor's Exam

📖 *NFPA 13 - Installation of Sprinkler Systems,* 1985, National Fire Protection Association, 1 Batterymarch Park, P. O. Box 9101, Quincy, MA 02269

📖 *NFPA 14 - Installation of Standpipe and Hose Systems,* 1993, National Fire Protection Association, 1 Batterymarch Park, P. O. Box 9101, Quincy, MA 02269

Recommended Reading for the Contractor's Trade Exam (continued)

Sprinklers and Fire Protection Contractor's Exam (continued)

📖 *NFPA 17 - Dry Chemical Extinguishing Systems,* 1990, National Fire Protection Association, 1 Batterymarch Park, P. O. Box 9101, Quincy, MA 02269

📖 *NFPA 17A - Wet Chemical Extinguishing Systems,* 1990, National Fire Protection Association, 1 Batterymarch Park, P. O. Box 9101, Quincy, MA 02269

📖 *NFPA 20 - Standard for the Installation of Centrifugal Fire Pumps,* 1990, National Fire Protection Association, 1 Batterymarch Park, P. O. Box 9101, Quincy, MA 02269

📖 *NFPA 24 - Private Fire Service Mains and Their Appurtenances,* 1992, National Fire Protection Association, 1 Batterymarch Park, P. O. Box 9101, Quincy, MA 02269

📖 *NFPA 96 - Standard for Ventilation Control and Fire Protection of Commercial Cooking Operations,* 1994, National Fire Protection Association, 1 Batterymarch Park, Box 9101, Quincy, MA 02269-9101

📖 *NFPA 101- Life Safety Code,* 1991, National Fire Protection Association, 1 Batterymarch Park, P. O. Box 9101, Quincy, MA 02269

📖 *NFPA 170- Fire Safety Symbols,* 1991, National Fire Protection Association, 1 Batterymarch Park, P. O. Box 9101, Quincy, MA 02269

📖 *NFPA 231- General Storage,* 1990, National Fire Protection Association, 1 Batterymarch Park, P. O. Box 9101, Quincy, MA 02269

📖 *NFPA 231C- Rack Storage of Materials,* 1991, National Fire Protection Association, 1 Batterymarch Park, P. O. Box 9101, Quincy, MA 02269

Electrical Contractor's Exam

📖 *Alternating Current Fundamentals,* Duff and Herman, Delmar Publishers, P. O. Box 6904, Florence, KY 41022

📖 *American Electricians Handbook,* 1996, 13th edition, Croft/Summers, McGraw-Hill Inc., Box 543, Blacklick, OH 43004-0543

📖 *Direct Current Fundamentals,* Loper and Tedsen, Delmar Publishers, P. O. Box 6904, Florence, KY 41022

📖 *NFPA 70 - National Electrical Code,* 1996 edition, National Fire Protection Association, 1 Batterymarch Park, P. O. Box 9101, Quincy, MA 02269

Department of Transportation (DOT)

To bid on any contract let by the West Virginia Division of Highways, you must get a Certificate of Qualification. To apply for this certificate, contact:

West Virginia Department of Transportation
Division of Highways — Construction Division
Building 5 Capitol Complex
1900 Kanawha Blvd. East
Charleston, WV 25305
(304) 558-2874
Fax (304) 558-2815

You'll be asked to fill out a Contractor's Prequalification Statement. It will ask you for:

- a financial statement
- what construction equipment you own
- what your work experience is
- what work you've completed
- what classes of work you want to be certified in

West Virginia issues certificates in these fields:

General construction	Pile driving, drilling
Signing	Tunneling
Portland cement concrete paving	Buildings
Sawing, sealing, curing, joint repair	Guardrail and fence
Bituminous paving	Grooving
Demolition	Seeding, sodding & mulching
Base courses, treated & untreated soil cement stabilization	Curb gutter, sidewalk, inlets, manholes & concrete medians
Cleaning & painting bridges	Landscaping
Grading	Miscellaneous projects (estimate cost less than $200,000)
Pavement markings	
Bridge construction	Traffic signals, electrical installations & lighting
Waterlines & sewers	
Drainage structures & culverts	

The Department will use this information to figure out how much you can be prequalified for. They'll round the amount to the nearest thousand dollars.

Once you get a Certificate of Qualification, it's good for no more than 16 months. The actual time will depend on your fiscal period and the date of the financial statement you submit on your Prequalification Statement.

Out-of-State Corporations

If your company isn't incorporated under the laws of West Virginia, you have to get Certificate of Authority from the Secretary of State to do business in West Virginia. Contact the Secretary of State at:

Office of Secretary of State
Building 1, Room W-139
1900 Kanawha Blvd. East
Charleston, WV 25305
(304) 558-8000
Fax (304) 558-0900

Wisconsin

To file a construction permit for a one- or two-family dwelling, you have to have a credential. To apply for the credential, contact:

Contact

Safety and Buildings Division
201 E. Washington Avenue
P. O. Box 7969
Madison, WI 53707-7969
(608) 261-5800

You will need to post a $25,000 bond or document $250,000 of liability insurance. You also have to document workers' compensation insurance and unemployment insurance. It will cost you $10 nonrefundable to file the application. The credential is $30 and it's good for one year.

Asbestos Abatement Certification

With a few exceptions, you need to be certified to work on asbestos abatement in Wisconsin. To apply for certification, contact:

Contact

The Asbestos Unit
Bureau of Public Health
1414 E. Washington Ave., Room 117
Madison, WI 53703-3044
(608) 261-8366

The Bureau issues the following types of certificates:

Worker
Supervisor
Inspector
Management planner

Project designer
Roofing worker
Roofing supervisor

To qualify for a certificate you must complete a training course and pass a closed book, multiple choice exam. Here's a summary showing each type of certificate, the length of its training course, number of questions on its exam, and the fee you have to pay for the certificate:

Worker

Length of training: 4 days

Number of exam questions: 50

Fee: $50

Supervisor

Length of training: 5 days

Number of exam questions: 100

Fee: $100

Inspector

Length of training: 3 days

Number of exam questions: 50

Fee: $150

Project designer

Length of training: 3 days

Number of exam questions: 100

Fee: $150

Management planner

Length of training: 3-day inspector course plus 2-day management planner course

Number of exam questions: 50

Fee: $100

Roofing worker

Length of training: 1 day

Number of exam questions: 35

Fee: $25

Roofing supervisor

Length of training: 1-day roofing worker course plus 1-day roofing supervisor course

Number of exam questions: 50

Fee: $50

If you have a valid certificate in another state and you got your certificate by completing an EPA-accredited training course or a course equal to Wisconsin's, you can petition the Bureau to be certified in Wisconsin. You won't have to repeat the training or work experience the Bureau requires. However, you may have to pass a Wisconsin exam and you'll still have to pay the certification fee.

A certificate is good for one year.

Lead Abatement Certification

With a few exceptions, you need to be certified to work on lead abatement in Wisconsin. To apply for certification, contact:

Contact

The Lead Unit
Bureau of Public Health
1414 E. Washington Ave., Room 167
Madison, WI 53703-3044
(608) 261-6876

The Bureau issues the following types of certificates:

Worker

Supervisor

Inspector

Risk assessor

Project designer

To qualify for a certificate you must complete a training course and pass a closed book exam. You have to have taken the training within 24 months or less from when you apply for a certificate. If more than 24 months have elapsed, you'll have to take a refresher course before you can get certified. If you took your training in another state, you can possibly get it approved by Wisconsin. Here's a summary showing each type of certificate, its training course, and the fee you have to pay for the certificate:

Worker
Training course(s) required: Worker
Fee: $50

Supervisor
Training course(s) required: Supervisor
Fee: $100

Inspector
Training course(s) required: Inspector
Fee: $150

Project designer

Training course(s) required: Project designer plus supervisor

Fee: $250

Risk assessor

Training course(s) required: Risk assessor plus inspector

Fee: $250

There are also some special requirements for supervisor, project designer, and risk assessor certificates. If you want to get a supervisor certificate you'll have to complete one of the following:

- one year of experience as a certified lead abatement worker
- two years of work experience in a related field such as asbestos, environmental remediation, or construction

For a project designer certificate you'll have to complete one of the following:

- bachelor's degree in engineering, architecture, or a related profession and one year of experience in building construction and design or a related field
- four years of work experience in building construction and design or a related field

For a risk assessor certificate you'll have to complete one of the following:

- bachelor's degree and one year of experience in building construction and design or a related field
- associate's degree and two years of experience in building construction and design or a related field
- high school diploma or equivalent and three years of experience in building construction and design or a related field
- hold a professional certification as an industrial hygienist, professional engineer, registered architect, safety professional, or environmental scientist

A certificate is good for one year.

Electrician's Credentials

In Wisconsin you can get a state master, journeyman, or electrical contractor credential which allows you to do electrical work in any municipality in the state that requires a license. To apply, contact:

Contact

Safety and Buildings Division
201 E. Washington Avenue
P. O. Box 7969
Madison, WI 53707-7969
(608) 261-8500

To get an electrical contractor's credential, you must be the head of your company and give the Division the following:

- your social security number
- your workers' compensation number
- your unemployment insurance account number
- your state tax identification number
- your federal tax identification number
- the names and addresses of all the officers of your company

Electrical contractor's credential fees: It'll cost you $35 nonrefundable to file an application. The credential costs $150 and it's good for three years.

To be certified as a master or journeyman you need to pass an open book exam on the National Electrical Code and the electrical code part of the Wisconsin Administrative Code. To qualify for the master electrician exam you need 1,000 hours per year of electrical construction work experience for at least seven years. You can substitute 500 hours of work experience for each semester you've completed in accredited electrical study, up to 3,000 hours.

There are two ways to qualify for the journeyman electrician exam. One way is to complete 1,000 hours per year of electrical construction work experience for at least five years, substituting 500 hours of work experience for each semester you've completed in accredited electrical study, up to 2,000 hours. The other way is to complete an electrical apprenticeship program recognized by Wisconsin and the U.S. Department of Labor.

Electrician's credential fees: It will cost you $65 nonrefundable to file the application and take the exam for master or journeyman. When you pass the exam you'll have to pay $50 to get your credential. The credential is good for three years.

Plumber's Credentials

To do plumbing work in Wisconsin you must get a credential. To apply for the credential, contact:

Safety and Buildings Division
201 E. Washington Avenue
P. O. Box 7969
Madison, WI 53707-7969
(608) 261-5800

The Division issues the following plumbing credentials:

Master	Journeyman - restricted appliances
Master - restricted appliances	Journeyman - restricted sewer services
Master - restricted sewer services	Utility contractor
Journeyman	Pipe layer

To be certified as a master or journeyman, you need to pass an open book exam on Chapter 145 of the Wisconsin Statutes and sections of the Wisconsin Administrative Code. You'll get a copy of Chapter 145 with your credential application.

To qualify for the master plumbing exam, you need 1,000 hours per year of work as a licensed journeyman plumber for at least three consecutive years or a degree in civil engineering, mechanical engineering, or other approved engineering degree related to plumbing. For either master plumber restricted exam, you need only two years of work experience as a journeyman.

To qualify for the journeyman exam you must complete a Division-approved plumbing apprenticeship program. For the journeyman (restricted appliances) exam you need 1,000 hours of work experience and the following educational courses:

- plumbing code - 40 hours
- blueprint reading - 20 hours
- transit or builder's level - 10 hours
- construction related mathematics - 20 hours
- first aid and safety - 10 hours

For the journeyman (restricted sewer services) you need 1,000 of work experience and these educational courses:

- plumbing code - 40 hours
- blueprint reading - 20 hours
- plumbing related mathematics - 10 hours
- appliance and equipment servicing - 30 hours

Plumber's credential fees: It will cost you $50 nonrefundable to file any master plumber application and take the exam. When you pass the exam you'll have to pay $250 to get your credential. The credential is good for two years.

It costs $30 nonrefundable to file any journeyman plumber application and take the exam. When you pass the exam you'll have to pay $90 to get your credential. The credential is good for two years.

To work on water service lines and sewers you need a utility contractor's credential To qualify for the utility contractor exam you have to be at least 18 years old. It costs $40 nonrefundable to file an application and take the exam. The credential costs $250 and it's good for two years.

To work on water service lines and sewers under the supervision of a licensed utility contractor, licensed master plumber, or a licensed master plumber (restricted sewer service), you need a pipe layer's credential. It'll cost $10 nonrefundable to file an application. The credential costs $90 and it's good for two years.

HVAC Credentials

To do HVAC work in Wisconsin you must get a credential. To apply for the credential, contact:

Safety and Buildings Division
201 E. Washington Avenue
P. O. Box 7969
Madison, WI 53707-7969
(608) 261-5800

HVAC credential fees: The Division issues HVAC contractor and qualifier credentials. To qualify for an HVAC contractor credential you must be the head of your company. It'll cost you $10 nonrefundable to file an application for a contractor credential. The credential costs $50 and it's good for two years.

To qualify for the HVAC qualifier credential exam you need to have 1,000 hours per year for four years of work experience in HVAC. You also need four years in an accredited technical school or a combination of work experience and study for four years.

It costs $30 nonrefundable to file an HVAC qualifier application and take the exam. When you pass the exam you'll have to pay $30 to get your credential. The credential is good for two years.

Fire Sprinklers Credentials

To work on fire sprinklers in Wisconsin you must get a credential. To apply for the credential, contact:

Safety and Buildings Division
201 E. Washington Avenue
P. O. Box 7969
Madison, WI 53707-7969
(608) 261-5800

The Division issues fire sprinkler contractor, journeyman fitter, sprinkler maintenance contractor, and sprinkler maintenance fitter credentials.

Fire sprinklers credential fees: To qualify for a fire sprinkler contractor credential you have to pass an exam. It costs $125 nonrefundable to file an application and take the exam. The credential costs $1000 and it's good for two years.

To qualify for the journeyman sprinkler fitter credential exam you need to have completed a Division-approved automatic fire sprinkler system apprenticeship. It will cost you $30 nonrefundable to file the application and take the exam. When you pass the exam you'll have to pay $90 to get your credential. The credential is good for two years.

To qualify for the sprinkler maintenance contractor credential you have to pass an exam. It costs $75 nonrefundable to file the application and take the exam. When you pass the exam you'll have to pay $200 to get your credential. The credential is good for two years.

To get a sprinkler maintenance fitter credential you have to file an application with the Division. It will cost you $10 nonrefundable to file the application and $30 to get your credential. The credential is good for two years.

Tank Installer's Credentials

To work on tanks in Wisconsin you must get a credential. To apply for a credential, contact:

Contact

Safety and Buildings Division
201 E. Washington Avenue
P. O. Box 7969
Madison, WI 53707-7969
(608) 261-5800

The Division issues tank specialty, aboveground tank system installer, underground tank system installer, tank system liners, and tank system removers and cleaners credentials.

Fee

Tank installer's credential fees: To qualify for a tank specialty firm credential you must be the head of your company. It will cost you $20 nonrefundable to file the application. The credential costs $50 and it's good for two years.

To qualify for any of the other credentials you need to pass an exam. It costs $20 nonrefundable to file the application and take the exam. When you pass the exam you'll have to pay $50 to get your credential. The credential is good for two years.

Pump Installer's Credentials

To do pump installation business in Wisconsin you must be registered with the Department of Natural Resources. To get an application, contact:

Contact

Department of Natural Resources
101 South Webster Street
P. O. Box 7921
Madison, WI 53707-7921
(608) 266-2621
Fax: (608) 267-3579

You must pass an exam the Department gives to prove you're competent to do pump installation work. The exam is on Department rules, well location and pump installation requirements, driven point well construction, and sampling and reporting requirements. The Department will send you a study guide for the exam after they get your completed application.

Pump installer's credential fees: You will have to pay a $25 application fee but not until the Department requests it. Then there's a $25 registration fee when you pass the exam and get your registration. Registration is good for one year and it expires on December 31 each year.

Well Driller's Credentials

To do well drilling business in Wisconsin you must be registered with the Department of Natural Resources. To get an application, contact:

Department of Natural Resources
101 South Webster Street
P. O. Box 7921
Madison, WI 53707-7921
(608) 266-2621
Fax: (608) 267-3579

You must pass an exam the Department requires before you can be registered. The exam is on Department rules, well construction, reconstruction, well abandonment, location requirements, and sampling and reporting requirements. To take the exam you must have two years of supervised well drilling experience within the last five years. You must also have drilled at least 30 wells or have 1,500 hours of well drilling in two years, with at least ten wells or 750 hours in a single year.

If you have a valid well drilling license in another state which you got by fulfilling requirements similar to Wisconsin's, you may qualify for the exam directly. You will need to submit Department Form 3300-94 and a photograph of yourself operating a drilling rig with your application. The Department will evaluate your license and notify you if you qualify or not.

The Department will send you a study guide for the exam after they get your completed application.

Well driller credential fees: You'll have to pay a $50 application fee but not until the Department requests it. Then there's a $50 registration fee when you pass the exam and get your registration. Registration is good for one year and it expires on December 31 each year.

Department of Transportation (DOT)

To bid on any Wisconsin Department of Transportation construction project, you must be prequalified by the Department. To get the Prequalification Statement, contact:

Contact

Wisconsin Department of Transportation
Bureau of Highway Construction
4802 Sheboygan Ave., Room 601
P. O. Box 7916
Madison, WI 53707-7916
(608) 266-1631

The Transportation Department will ask you for these items:

- financial statement
- which types of work you want to be prequalified for
- your current and past work projects
- what equipment your firm owns
- maximum amount of work (in dollars) you would be willing to undertake

You'll also be asked which of the following types of work you want to be prequalified in and what the maximum amount of work (in dollars) you think you can do for each of those types:

General construction	Rail construction or rehabilitation
Grading	Bridge painting
Concrete pavement	Street or airport lighting
Asphaltic pavement	Building construction
Gravel or crushed stone	Incidental construction
Structures	

The Department will use this information to give your company a prequalification rating. Prequalification is good for 16 months from the date of your financial statement.

Out-of-State Corporations

Out-of-state corporations doing business in Wisconsin must qualify with the Wisconsin Secretary of State to do business in the state. For information, contact:

Contact

Corporation Division
Office of the Secretary of State
P. O. Box 7846
Madison, WI 53707
(608) 266-3590
Fax (608) 267-6813

Wyoming

All contractors, except electrical, are licensed at the local (city or county) level. However, the state requires everyone doing electrical work in Wyoming to be licensed.

Electrician's Licenses

If you work as an apprentice electrician, you need to be registered in Wyoming. Here are the electrical licenses Wyoming issues:

Contractor	Journeyman
Master	Low voltage/limited (contractor and technician)

To get an application for a license, contact:

Contact

State of Wyoming — Electrical Board
Department of Fire Prevention & Electrical Safety
Cheyenne, WY 82002
(307) 777-7991
Fax (307) 777-7119
E-mail Bholli@Missc.State.WY.US

Electrical Contractor

You don't have to take an exam to get an electrical contractor's license but you must agree that all the work you do will be under the direction and supervision of a master electrician that you employ. Furthermore, all your electrical employees must be licensed masters, licensed journeymen, journeymen who hold valid temporary work permits, or registered apprentice electricians.

Fee

Electrical contractor's license fees: An electrical contractor's license costs $200 and it expires every year on July 1. Renewing the license will cost you $100.

Master Electrician

To apply for a master electrician's license you need eight years (16,000 hours) of experience in residential, commercial, and industrial electrical work. You can apply up to two years (4,000 hours) study in electrical courses to the work experience requirement.

After the Board checks your qualifications and decides you're eligible to take the exam they require for the license, they'll send you an Exam Registration Form. You'll need a whole day for the exam. The three-hour morning session has questions on electrical theory, the code, and the Wyoming State electrical rules and regulations. The four-hour afternoon session is made up of code calculation questions. Here are the subjects covered and their percentages followed by a reference list:

Subject	Percent of exam
Grounding and bonding	11
Services, feeders, branch circuits, and overcurrent protection	12
Raceways and enclosures	10
Conductors	9
Motor and controls	11
Utilization and general use equipment	11
Special occupancies/equipment	6
General knowledge of the electrical trade and calculations	25
Low voltage circuits including alarms and communications	3
State laws	2

Master electrician's license fees: It costs $25 to file the application for a master's license, $50 for the exam, and $100 to get the license. It's good until July 1 of the third year after you got it.

Wyoming recognizes Utah and South Dakota master electrician licenses. It also has reciprocal agreements with some other states. Contact the Board for their policy with your home state.

Journeyman Electrician

To apply for a journeyman's license you need to have four years (8,000 hours) of apprenticeship or four years (8,000 hours) of work experience in residential, commercial, and industrial electrical wiring work. You can apply up to two years (4,000 hours) of study in electrical courses to the work experience requirement. Then the process is the same as for the master's license.

However the journeyman exam isn't quite the same as the master exam. It has a practical part to test your mechanical skills, which lasts a bit less than an hour, and a written part that's similar to the master exam. This part is open book, multiple choice, and lasts three hours. Here are the subjects covered in the written exam and their percentages:

Subject	Percent of exam
Grounding and bonding	11
Services, feeders, branch circuits, and overcurrent protection	14
Raceways and enclosures	14
Conductors	12
Motor and controls	10
Utilization and general use equipment	10
Special occupancies/equipment	6
General knowledge of the electrical trade and calculations	18
Low voltage circuits including alarms and communications	3
State laws	2

Journeyman electrician's license fees: It will cost you $25 to file an application for a journeyman's license. The written and practical exams cost $75 if you take both of them, or $50 each if you take them separately. If you have a current journeyman or master license from another state you only have to take the written exam. The license costs $50 and it's good until January 1 of the third year after you got it.

If you have had a journeyman license from Utah, Idaho, Montana, South Dakota, Oregon, Washington, or Alaska for at least a year and took an exam to get it, Wyoming may recognize your license.

Low Voltage/Limited Electrical Contractor

You don't have to take an exam to get a low voltage/limited electrical contractor's license but you must agree that all your employees are low voltage/limited technicians, registered apprentice technicians, or hold valid temporary low voltage/limited technician work permits.

Low voltage/limited electrical contractor's license fees: A low voltage/limited electrical contractor's license costs $100 and it expires every year on July 1. Renewing the license will cost you $50.

Limited Technician

Wyoming issues limited technician's licenses in these trades:

Elevator

Electric signs

Well water and irrigation

Light fixtures

Heating, ventilating, and air conditioning systems

To apply for a limited technician license you need to have two years (4,000 hours) of apprenticeship or two years (4,000 hours) work of experience in one of these trades. You can apply up to two years (4,000 hours) of study in electrical courses to the work experience requirement.

The limited technician exam is open book, multiple choice format and lasts two hours. It's on:

- grounding and bonding
- services, feeders, branch circuits, and overcurrent protection
- raceways and enclosures
- conductors
- motors and controls
- utilization and general use equipment
- special occupancies/equipment
- general knowledge of the electrical trade and calculations
- low voltage circuits including alarms and communications
- state laws
- electronic components
- power supplies
- control circuits

Limited technician license fees: It will cost you $25 to file an application for a limited technician license. The exam costs $50. The license costs $50 and it's good until 1 July of the third year after you got it.

Low Voltage Technician

Wyoming issues low voltage technician's licenses in these trades:

General

Alarms

Communications

Sound

Television

Control

To apply for a low voltage technician license you need to have two years (4,000 hours) of apprenticeship or two years (4,000 hours) of work experience in one of these trades. You can apply up to two years (4,000 hours) of study in electrical courses to the work experience requirement.

The low voltage technician exam is open book, multiple choice format and lasts two hours. The subjects covered on the test are:

- branch circuits
- conductors
- equipment
- electronic components
- power supplies
- grounding and bonding
- general knowledge
- raceways and enclosures
- state law
- overcurrent protection
- sound and communications
- special occupancies
- fire alarm systems
- burglar alarm
- control circuits

Here's an outline that tells you the percentage of questions there are for each subject by trade:

Subject	Low voltage technician license trades					
	General	Alarms	Comm.	Sound	TV	Control
Branch circuits	12	12	12	12	12	12
Conductors	12	12	12	12	12	12
Equipment	6	10	12	12	12	12
Electronic components	10	10	11	11	11	11
Power supplies	10	10	10	10	10	10
Grounding and bonding	10	10	10	10	10	10
General knowledge	6	6	10	10	10	8
Raceways and enclosures	2	4	6	6	6	6
State law	2	2	3	3	3	3
Overcurrent protection	2	2	2	2	2	2
Sound and communications	10	—	12	12	12	—
Special occupancies	2	2	—	—	—	2
Fire alarm systems	8	10	—	—	—	—
Burglar alarm	8	10	—	—	—	—
Control circuits	—	—	—	—	—	12

Fee

Low voltage technician license fees: It will cost you $25 to file an application for a low voltage technician license. The exam costs $50. The license costs $50 and it's good until 1 July of the third year after you got it.

Department of Transportation (DOT)

To bid on any Wyoming Department of Transportation construction project, you must be prequalified by the Department. To get the Prequalification Questionnaire, contact:

Prequalification Officer
Construction & Maintenance Division
Cheyenne, WY 82003-1708
(307) 777-4057

The Transportation Department will ask you for these items:

- financial statement
- which types of work you want to be prequalified for
- disadvantaged business status of your firm
- equal employment opportunity affidavit
- your current and past work projects
- what equipment your firm owns

You'll also be asked which of the following types of work you want to be prequalified in:

Grading	Landscaping
Gravel, asphalt, or concrete surfacing	Electrical work
Bridges	Underground utilities
General building construction (frame, metal, or masonry)	General
Guardrails and fences	Consulting
Signs	Other

The Department will use this information to give your company a prequalification rating. Prequalification is good for 15 months.

Out-of-State Corporations

Out-of-state corporations doing business in Wyoming must register with the Wyoming Secretary of State. For information, contact:

Corporations and UCC Department
Office of the Secretary of State
110 Capitol Building
Cheyenne, WY 82002
(307) 777-7311
Fax (307) 777-5339

Practical References for Builders

CD Estimator

If your computer has *Windows*™ and a CD-ROM drive, *CD Estimator* puts at your fingertips 85,000 construction costs for new construction, remodeling, renovation & insurance repair, electrical, plumbing, HVAC and painting. You'll also have the *National Estimator* program — a stand-alone estimating program for *Windows*™ that *Remodeling* magazine called a "computer wiz." Quarterly cost updates are available at no charge on the Internet. To help you create professional-looking estimates, the disk includes over 40 construction estimating and bidding forms in a format that's perfect for nearly any word processing or spreadsheet program for *Windows*™. And to top it off, a 70-minute interactive video teaches you how to use this CD-ROM to estimate construction costs. **CD Estimator is $68.50**

National Construction Estimator

Current building costs for residential, commercial, and industrial construction. Estimated prices for every common building material. Provides man-hours, recommended crew, and gives the labor cost for installation. Includes a CD-ROM with an electronic version of the book with *National Estimator*, a stand-alone *Windows*™ estimating program, plus an interactive multimedia video that shows how to use the disk to compile construction cost estimates. **560 pages, 8¹/₂ x 11, $47.50. Revised annually**

Excavation & Grading Handbook Revised

Explains how to handle all excavation, grading, compaction, paving and pipeline work: setting cut and fill stakes (with bubble and laser levels), working in rock, unsuitable material or mud, passing compaction tests, trenching around utility lines, setting grade pins and string line, removing or laying asphaltic concrete, widening roads, cutting channels, installing water, sewer, and drainage pipe. This is the completely revised edition of the popular guide used by over 25,000 excavation contractors. **384 pages, 5¹/₂ x 8¹/₂, $22.75**

Residential Steel Framing Guide

Steel is stronger and lighter than wood — straight walls are guaranteed — steel framing will not wrap, shrink, split, swell, bow, or rot. Here you'll find full page schematics and details that show how steel is connected in just about all residential framing work. You won't find lengthy explanations here on how to run your business, or even how to do the work. What you will find are over 150 easy-to-ready full-page details on how to construct steel-framed floors, roofs, interior and exterior walls, bridging, blocking, and reinforcing for all residential construction. Also includes recommended fasteners and their applications, and fastening schedules for attaching every type of steel framing member to steel as well as wood. **170 pages, 8¹/₂ x 11, $38.80**

Contractor's Guide to the Building Code Revised

This new edition was written in collaboration with the International Conference of Building Officials, writers of the code. It explains in plain English exactly what the latest edition of the *Uniform Building Code* requires. Based on the 1997 code, it explains the changes and what they mean for the builder. Also covers the *Uniform Mechanical Code* and the *Uniform Plumbing Code*. Shows how to design and construct residential and light commercial buildings that'll pass inspection the first time. Suggests how to work with an inspector to minimize construction costs, what common building shortcuts are likely to be cited, and where exceptions may be granted. **368 pages, 8¹/₂ x 11, $39.00**

How to Succeed With Your Own Construction Business

Everything you need to start your own construction business: setting up the paperwork, finding the work, advertising, using contracts, dealing with lenders, estimating, scheduling, finding and keeping good employees, keeping the books, and coping with success. If you're considering starting your own construction business, all the knowledge, tips, and blank forms you need are here. **336 pages, 8¹/₂ x 11, $24.25**

Building Contractor's Exam Preparation Guide

Passing today's contractor's exams can be a major task. This book shows you how to study, how questions are likely to be worded, and the kinds of choices usually given for answers. Includes sample questions from actual state, county, and city examinations, plus a sample exam to practice on. This book isn't a substitute for the study material that your testing board recommends, but it will help prepare you for the types of questions — and their correct answers — that are likely to appear on the actual exam. Knowing how to answer these questions, as well as what to expect from the exam, can greatly increase your chances of passing. **320 pages, 8¹/₂ x 11, $35.00**

Construction Forms & Contracts

125 forms you can copy and use — or load into your computer (from the FREE disk enclosed). Then you can customize the forms to fit your company, fill them out, and print. Loads into *Word for Windows, Lotus 1-2-3, WordPerfect, Works,* or *Excel* programs. You'll find forms covering accounting, estimating, fieldwork, contracts, and general office. Each form comes with complete instructions on when to use it and how to fill it out. These forms were designed, tested and used by contractors, and will help keep your business organized, profitable and out of legal, accounting and collection troubles. Includes a CD-ROM for *Windows*™ and Macintosh. **400 pages, 8¹/₂ x 11, $39.75**

Contractor's Index to the 1997 *Uniform Building Code*

Finally, there's a common-sense index that helps you quickly and easily find the section you're looking for in the *UBC*. It lists topics under the names builders actually use in construction. Best of all, it gives the full section number and the actual page in the *UBC* where you'll find it. If you need to know the requirements for windows in exit access corridor walls, just look under Windows. You'll find the requirements you need are in Section 1004.3.4.3.2.2 in the *UBC* — on page 115. This practical index was written by a former builder and building inspector who knows the *UBC* from both perspectives. If you hate to spend valuable time hunting through pages of fine print for the information you need, this is the book for you. **192 pages, 8¹/₂ x 11, $26.00. Loose-leaf edition, $29.00.**

Builder's Guide to Accounting Revised

Step-by-step, easy-to-follow guidelines for setting up and maintaining records for your building business. This practical, newly-revised guide to all accounting methods shows how to meet state and federal accounting requirements, explains the new depreciation rules, and describes how the Tax Reform Act can affect the way you keep records. Full of charts, diagrams, simple directions and examples, to help you keep track of where your money is going. Recommended reading for many state contractor's exams. **320 pages, 8¹/₂ x 11, $26.50**

Basic Engineering for Builders

If you've ever been stumped by an engineering problem on the job, yet wanted to avoid the expense of hiring a qualified engineer, you should have this book. Here you'll find engineering principles explained in non-technical language and practical methods for applying them on the job. With the help of this book you'll be able to understand engineering functions in the plans and how to meet the requirements, how to get permits issued without the help of an engineer, and anticipate requirements for concrete, steel, wood and masonry. See why you sometimes have to hire an engineer and what you can undertake yourself: surveying, concrete, lumber loads and stresses, steel, masonry, plumbing, and HVAC systems. This book is designed to help the builder save money by understanding engineering principles that you can incorporate into the jobs you bid. **400 pages, 8¹/₂ x 11, $34.00**

Quicken for Contractors

Most builders, contractors, and remodelers came up through the ranks, learning their craft "in the trenches." They know how buildings go together, but office management skills like bookkeeping, accounting, payroll and job costing are often a new, and dangerous, challenge. *Quicken for Contractors* explains how to use *Quicken*, an affordable, easy-to-understand computer program published by Intuit. It shows step-by-step how *Quicken* can work in the builder's office, putting accurate bookkeeping within easy reach. This manual does not include the program, but has instructions and examples, with onscreen "pictures" of familiar forms like checks, deposit slips and time cards that you can create with *Quicken*. Even if you only have a few minutes a week, this book will help you use *Quicken* to set up a basic checkbook system that will quickly tell you your cash position and job costs. You'll learn how to set up your financial files and create valuable reports, including profit & loss statements, payroll reports and job cost reports. The companion diskette included with the book consists of a sample construction company that you can use as a tutorial, or as a template for you to plug in your own data. **240 pages, 8¹/₂ x 11, $32.50**

Electrician's Exam Preparation Guide

Need help in passing the apprentice, journeyman, or master electrician's exam? This is a book of questions and answers based on actual electrician's exams over the last few years. Almost a thousand multiple-choice questions — exactly the type you'll find on the exam — cover every area of electrical installation: electrical drawings, services and systems, transformers, capacitors, distribution equipment, branch circuits, feeders, calculations, measuring and testing, and more. It gives you the correct answer, an explanation, and where to find it in the latest NEC. Also tells how to apply for the test, how best to study, and what to expect on examination day. **352 pages, 81/2 x 11, $28.00**

Blueprint Reading for the Building Trades

How to read and understand construction documents, blueprints, and schedules. Includes layouts of structural, mechanical, HVAC and electrical drawings. Shows how to interpret sectional views, follow diagrams and schematics, and covers common problems with construction specifications. **192 pages, 5¹/₂ x 8¹/₂, $14.75**

Craftsman's Illustrated Dictionary of Construction Terms

Almost everything you could possibly want to know about any word or technique in construction. Hundreds of up-to-date construction terms, materials, drawings and pictures with detailed, illustrated articles describing equipment and methods. Terms and techniques are explained or illustrated in vivid detail. Use this valuable reference to check spelling, find clear, concise definitions of construction terms used on plans and construction documents, or learn about little-known tools, equipment, tests and methods used in the building industry. It's all here. **416 pages, 8¹/₂ x 11, $36.00**

Basic Lumber Engineering for Builders

Beam and lumber requirements for many jobs aren't always clear, especially with changing building codes and lumber products. Most of the time you rely on your own "rules of thumb" when figuring spans or lumber engineering. This book can help you fill the gap between what you can find in the building code span tables and what you need to pay a certified engineer to do. With its large, clear illustrations and examples, this book shows you how to figure stresses for pre-engineered wood or wood structural members, how to calculate loads, and how to design your own girders, joists and beams. Included FREE with the book — an easy-to-use version of NorthBridge Software's *Wood Beam Sizing* program. **272 pages, 8¹/₂ x 11, $38.00**
